CANADA'S VICTORIAN OIL TOWN

Canada's Victorian Oil Town

The Transformation of Petrolia from a Resource Town into a Victorian Community

CHRISTINA BURR

McGill-Queen's University Press
Montreal & Kingston • London • Ithaca

ISBN-13: 978-0-7735-3112-3 ISBN-10: 0-7735-3112-2

Legal deposit fourth quarter 2006
Bibliothèque nationale du Québec

Printed in Canada on acid-free paper

This book has been published with the help of grants from the Canadian Federation for the Humanities and Social Sciences, through the Aid to Scholarly Publications Programme, using funds provided by the Social Sciences and Humanities Research Council of Canada, and from the University of Windsor.

McGill-Queen's University Press acknowledges the support of the Canada Council for the Arts for our publishing program. We also acknowledge the financial support of the Government of Canada through the Book Publishing Industry Development Program (BPIDP) for our publishing activities.

Library and Archives Canada Cataloguing in Publication

Burr, Christina Ann, 1959-
Canada's Victorian oil town : the transformation of Petrolia from a resource town into a Victorian community / Christina Burr.

Includes bibliographical references and index.
ISBN-13: 978-0-7735-3112-3
ISBN-10: 0-7735-3112-2

1. Petrolia (Ont.) – History. 2. Group identity – Ontario – Petrolia – History. 3. Community life – Ontario – Petrolia – History. 4. Petroleum industry and trade – Social aspects – Ontario – Petrolia – History. I. Title.

FC3099.P5B87 2006 971.3'27 C2006-902086-8

Typeset by Jay Tee Graphics Ltd. in New Baskerville 10/13

For Mom, Dad,
Simba, and Stuart

Contents

Acknowledgments

On 14 September 1887, twenty-one-year-old Christina Coulter, my great-grandmother, married Samuel Phillips, age twenty-five, of Petrolia, Ontario. Sam Phillips was an oil driller by occupation. He learned his trade locally, in the oil-rich fields of Enniskillen Township in Lambton County. The capriciousness of the oil industry meant that when the Canadian markets slumped local drillers were often unemployed. Some found work in other local industries, but the prospects of adventure for young men raised in rural Ontario and of good wages for men with families to support led many drillers to foreign fields. Sam Phillips spent much of his married life working abroad. When he died on 20 May 1919, at the age of 57, his obituary in the *Petrolia Advertiser-Topic* revealed that he had been a manager of a refinery in Germany for several years. He also worked in Africa, Java, Borneo, South America, Egypt, and Persia, being in the latter country when World War I broke out in 1914. With the exception of a three-year period from 1897 to 1900, when Christina accompanied her husband to Germany, she remained in the couple's house on Catherine Street in Petrolia, where she raised their two surviving children. It was during the couple's sojourn in Germany that they acquired a beautiful set of handmade china embossed in gold that my mother inherited. I grew up hearing small bits of the story of the foreign drillers at holiday time and on special occasions when the china was brought out for dinner. It is this personal story that sparked the research presented here, and I owe a debt of gratitude to many people and institutions that assisted me in tracing Enniskillen Township's oil heritage that I am delighted to acknowledge here.

For financial assistance at various stages in the research, I am indebted to the Social Sciences and Humanities Research Council of Canada for a Standard Research Grant and additional funding from the Aid to Scholarly Publications Programme to publish the book. The University of Windsor provided financial support in the form of a Research Grant for Women Faculty and funds for publication of this book. My greatest debts for archival assistance were incurred at the Lambton Room in the Lambton County Library in Wyoming, Ontario. Helen Maddock and Anne Ashton spent hours answering my questions, digging through archives boxes for sources, and providing insights from their own knowledge of Enniskillen Township's oil drillers that helped me to develop the arguments presented in this book. I am grateful not only for their knowledge of local history but also for the congenial environment they provided during my many weeks at the Lambton Room. I also want to acknowledge the assistance of the staffs at the Petrolia Discovery and the Oil Museum of Canada, who answered my questions and gave me tours of their museums. Special thanks also to Charlie Fairbank and Kathleen Gillespie, who gave me access to personal documents pertaining to their families' involvement in the oil industry and community life in Enniskillen Township. I am also grateful to the staffs at both Archives Canada and the Archives of Ontario for assistance in wading through the records of the Geological Survey of Canada and the Crown Lands Records.

At McGill-Queen's University Press I have had the good fortune to have the support of Roger Martin as my editor. Roger guided the manuscript through the lengthy review process and encouraged me to be patient. Lesley Andrassy's meticulous copy editing improved the manuscript immensely. My thanks also to Joan McGilvray for bringing it all together.

Much of chapter 1 was published as "'Oil Mania:' Colonial Land Policy, Land Speculation, and Settlement in Enniskillen Township, 1830s–1860s," in *Histoire sociale/Social History* (November 2005).

A version of chapter 5 was published as "'Some Adventures of the Boys': Enniskillen Township's 'Foreign Drillers,' Imperialism, and Colonial Discourse, 1873–1923," in *Labour/Le Travail* (Spring 2003). I thank the editors of both journals for their permission to use these works here.

My greatest debts are to my parents Barbara and Frank Burr. They toured the museums with me, helped arrange to have photographs reproduced, housed and fed me when I came "home" to do research, and listened to me as I tried to work through my ideas. They have supported my academic career all the way and have given me a rich family life. I am happy to acknowledge their contribution here.

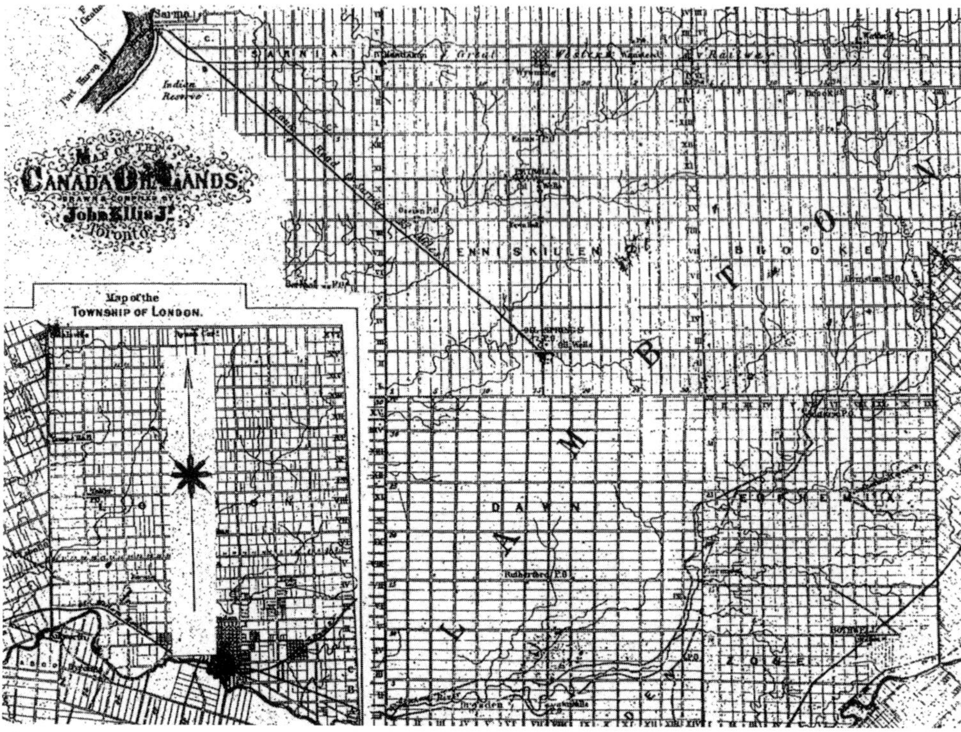

Map of Enniskillen Township. *Belden's Illustrated Historical Atlas of the County of Lambton*

CANADA'S VICTORIAN OIL TOWN

INTRODUCTION

Colonization, Culture, and The Making of Canada's Victorian Oil Town

Canada's Victorian Oil Town! Beginning in the 1970s, heritage initiatives in the small town of Petrolia, Ontario, by public agencies and community non-profit organizations resulted in the commercialization of the town's origins as a Victorian "oil boom town" to promote economic revitalization by building a tourist industry. This representation of Petrolia as an orderly, hard-working, law-abiding, and respectable Victorian community attests to the importance of history, memory, and place in the creation of imagined community. Ordinary landscapes and buildings, as well as Victoria Hall, an extraordinary local restoration initiative that now houses a municipal office and a lucrative summer theatre program, have been particularly effective in claiming the symbolic importance of place. The building of a community identity around the town's origins in the oil industry has been prompted by the restoration of the historic oil field called the Petrolia Discovery. A living museum was created using nineteenth-century oil drilling relics that have been made operational once again.

A community's collective values, beliefs, and practices are expressed through the creation of particular narratives of the past. What are the social, cultural, and historical processes by which Petrolia has come to be remembered as "Canada's Victorian Oil Town?" A little known story in Canadian history is that a thriving commercial oil industry was built in Enniskillen Township, in the centre of Lambton County, beginning in the early 1850s, nearly a century before oil was struck in Leduc, Alberta, in 1947. This book is about the cultural dimensions of

British colonization of Enniskillen Township, and the transformation of Petrolia from a rough shanty-town of disreputable land speculators and "wildcatters" into a representation of an orderly, respectable, and "civilized" community, which since the 1970s has been marketed for tourism as "Canada's Victorian Oil Town." This study contributes to the already extensive historiography of the role of resource development in Canadian society. In particular, I have gleaned many useful insights from those historians who focused on the working-class experience in resource communities with attention to the economies of resource extraction. The research presented here departs from those studies, however, with its attention to the shifting historical and ideological contexts of community-making.[1] Emphasis is placed on the cultural importance of place and on how in Petrolia the oil industry provides the expression and precise definition of cultural boundaries and community identification. My study has been influenced also by efforts to bring colonization back into the writing of Canadian history with a concern for the dialectic between the local and the global.[2] I suggest that British ruling-class ideals of structure and order were never simply imposed on the settler colony, but were shaped in a local social and cultural context around the construction of a community identity rooted in the local oil industry. Connections are made between science and technology, nation and Empire, and, more recently, the appropriation of heritage for tourism.

Lambton County forms the western frontier and the southern border with the U.S., fronted by the St Clair River, for a distance of sixty miles or more. Geographically Enniskillen Township constitutes the central township of Lambton County, bordered on the north by Plympton Township, on the east by Brooke, on the south by Dawn, and on the west by Moore and Sarnia Townships. In the aftermath of the War of 1812, colonial officials encouraged the immigration of persons of British origin who, in their opinion, would be more loyal than those settlers who had previously migrated from the United States. They sought land for British settlers in the Western District of Upper Canada, where U.S. incursions were most likely to occur.

The first step in the colonization of Enniskillen Township was the alienation of Aboriginal claims by the Crown. The Ojibwa (Chippewa) First Nations ceded the lands in what would become the two south concessions of Enniskillen Township in two treaties completed in 1822 and 1827. The Native peoples never relinquished ownership of the subsurface or submarine rights in their reserves, in their unceded lands, or in their territories ceded by treaty. Thus, the Crown never had title to the sub-surface and any of its resources, and still does not, even though most of the oil resources of Enniskillen Township have now been extracted. On 13 March 1841 Lewis Rendt sold the east half of lot nine, concession ten, for 37 pounds and ten shillings Canadian to the Crown. The following year, on 8 March 1842, David McCall sold 300 acres on the east half of lot eight, concession ten, and lot eight, concession nine, to the Crown for 220 pounds Canadian. The lands sold to the Crown were intended as reserves for the Chippewa Indians of the St Clair River and the Chenail Ecarté tribe, but instead were sold again during the height of the Enniskillen oil boom in the 1860s. The ownership of the property is the subject of an ongoing court case, with the Chippewa Indians of the Sarnia, Kettle Point, and Walpole Island bands arguing that the land was held in trust by the federal government for the bands.[3]

Chapter 1 begins with the work of Lewis Burwell, who, along with Eliakim Malcolm, surveyed Enniskillen Township between 3 September and 4 December 1832. As the extensive international literature on colonialism and cultural geographies suggests, nineteenth-century European practices of geographical reconnaissance made the colonies amenable to colonization.[4] Major transformations in the organization of colonial governance were initiated in the early decades of the nineteenth century in the settler colony of Upper Canada. The overall objective was to pair bureaucratic administrative centralization with local representative government by men of property and good moral character. The government of Upper Canada was transformed into an experiment in liberal political administration. The creation of an administrative infrastructure and an official document system was part of the process of colonization. As historian Bruce Curtis notes, "The centre of gravity of documentary production shifted increasingly towards the colonial state as the administrative capacities of the latter were extended and intensified."[5] The key figures in the administration of colonization were local agents and various experts, who produced an archive of maps, statistics, and registers. The work of the Surveying Branch of the Crown Lands Department was important to the administration of the settler colony. Surveyors created and defined the spatial order of the empire. They laid out the property lines and township grids, and they evaluated the potential of these lands for settlement.

The survey of Enniskillen Township was part of the power-knowledge processes whereby imperial claims to the territory were legalized and normalized using Victorian ideas of science and spatial order. British ideals of utilitarianism, as historian Suzanne Zeller and others have argued, laid the conceptual foundation for the spatial ordering of British North America.[6] The inventory science of botany was used to promote an economy based on agriculture and forestry. Nineteenth-century surveyors distinguished between "good" land for settlement and agriculture and "poor," "desert" land on the basis of tree species and vegetation.

In 1835 Enniskillen Township was opened up for settlement. Colonization, together with the implementation of British ideals of "improvement," was part of a concerted strategy for overcoming "the struggle with distance," especially in the frontier regions of the Western District. Tensions arose between colonial administrators and those acquiring land and sometimes settling in Enniskillen Township. Squatters and land speculators acquired and disposed of lands in ways that did not concur with imperial models of affiliation, loyalty, and dependence. Land policies in Upper Canada did not affect every township equally, however, as the studies by Clarke, Shannon, and Widdis have illustrated, particularly when such factors as physical geography, the timing and extent of European settlement, and absentee ownership are taken into consideration.[7] Colonialism was never a uniform or coherent imposition, but a practically mediated set of relationships. Imperial discourses were bent from colonial vantage points. Crown land agents were given considerable leeway. Most settlers dealt with the regulatory regime of the state at the local level. The local Crown land agent allocated land grants and managed the sale of lots. As Daniel Clayton writes, "colonial places have histories and geographies that exceed what the centre made of them in administrative and historiographic terms." Thus, what Clayton refers to as "the messy pragmatics" of settler colonialism must be unravelled in micro-studies of townships to fully comprehend the effects of colonial land policies on settlement.[8]

The countryside was transformed almost overnight in the summer of 1858, when James Miller Williams, a successful Hamilton carriage manufacturer and former alderman, dug the first commercially successful oil well on property he had acquired from Charles Nelson Tripp. The demand for petroleum for the manufacture of illuminating oil and as a lubricant for machinery increased over the latter half of the nineteenth century. Petroleum lubricants literally greased the machinery that powered the industrial revolution. Speculators, many of them from the U.S., poured into Enniskillen Township, and the "first oil boom" was underway, accompanied by the emergence of the village of Oil Springs. Throughout the early 1860s the land market in Enniskillen was exceedingly volatile. Property changed hands quickly, and the speculator in oil lands was typically viewed with suspicion.

Chapter 2 explains how colonization and capitalism intersected with science in the Enniskillen oil fields. Nineteenth-century travellers to the Western District recognized the medicinal uses of petroleum, which was sold to local merchants by the Native peoples and white settlers. They were also cognizant of eighteenth-century European geological theories about the common organic origins of coal and petroleum. Travel writers brought ideas about the earth sciences from their home culture, and imposed them on the new landscape in their efforts to establish order over the wilderness and provide a way of "seeing" the landscape that their readers at home in Britain could understand.

Geological surveys were productive ways for colonial governments to encourage the exploitation of natural resources without themselves going into the business of resource extraction. By studying the work of the geological survey in the Enniskillen oil fields, we can begin to flesh out how power was lodged in particular institutions and worked out locally in the colonial context. The production of scientific work involved the intersection of different "social worlds," and acquired its shape and meaning by virtue of the spatial, social, and cultural circumstances in which it was made and used. In doing so, I have been influenced by recent contributions to the historical writing on Victorian science and the revelation that professional scientists cooperated with local artisans and practical men in the field.[9] Geologists and practical men working in the field made contributions to the science of geology that were not fashioned solely in the interests of British imperialism.

A popular culture of science emerged around the first oil boom in Enniskillen Township in the midst of debates about the extent and supply of local oil resources, particularly in the aftermath of the widespread cessation of the surface flowing wells early in 1863. Local oilmen, by and large, did not receive professional lecturers with much enthusiasm, since they had not been able to pinpoint the exact location and extent of the local oil resources. Two popular books were published; one focussed exclusively on the oil fields of Enniskillen Township and the other drew on the oil fields to sketch a broader theory about the origins of creation. In 1865 Henry White, a land surveyor and mineral explorer, published a guidebook entitled *Geology, Oil Fields, and Minerals of Canada West: How and Where to Find Them,* intended for practical oilmen with only a limited knowledge of geology.[10] Alexander Winchell, a professor, popular writer, and lecturer, probably conducted fieldwork in Enniskillen Township during the 1860s while he was a professor at the University of Michigan. In *Sketches of Creation,* published in 1873, Winchell probed the relationship between science and religion that preoccupied the Victorians. He argued that petroleum, a beneficent provision from the earth's crust provided by God, was for man's use only.

The collapse of the first oil boom at Oil Springs in 1863, accompanied by a rise in the price of oil on the market, ushered in the second oil boom at Petrolia, approximately four miles south of the railway connection at Wyoming on the flat lands surrounding Bear Creek. Chapter 3 examines the creation of the oil resource town of Petrolia in the latter decades of the nineteenth century. Research in cultural geography by Edward Soja, Robert Shields, Anthony Cohen, Nigel Thrift, Doreen Massey, Henry Lefebvre, Michel de Certeau, and others has been helpful in explaining how communal boundaries are constructed in socio-cultural space.[11] Community is defined as an overlapping geographical, cultural, and acquaintance space, and boundaries are constructed and maintained through the discourses of everyday life. These community boundaries incorporate and enclose difference. "Insiders" shared access to a common discourse built around the vagaries of the oil industry. Prices on the oil exchange had an impact throughout the community and the Oil Exchange Hall emerged as a focal point of the community. In other words, as I discovered, place matters. It provides the economic, physical, and cultural context for class formation and the construction of a local community identity.

As the oil economy of Petrolia was integrated more extensively into national and international trading patterns in the early 1870s, the material conditions were present for the emergence of a middle class in Petrolia. The experience of class, as historians of middle-class formation have illustrated, was not the same for men and women. The ideology of separate spheres, as Mary Ryan, Catherine Hall, and Leonore Davidoff argue, was integral to the construction of middle-class identity.[12] "Separate spheres," a powerful nineteenth-century prescriptive ideology, assigned to men active and dominant roles in the public sphere of business and politics and to women the nurturing domestic roles in the private sphere of the home. The male middle class of wealthy "oil barons," merchants, manufacturers, and professionals set about to transform Petrolia from a rough-cast mining town into a public space that was orderly, regulated, and "civilized."

Chapter 4 is an analysis of the extensive male associational life built in Petrolia around a network of fraternal orders, fire brigades, the town band, and amateur sports clubs. It provides insight into how public space was constructed as a middle-class masculine terrain. Public space was constructed primarily on the basis of exclusions, notably, those of class, gender, ethnicity, and religion, from which the Anglo-Protestant male middle class derived its power. Friendly societies inculcated middle-class ideals of masculinity, including sobriety, thrift, piety, self-restraint, and moral obligation. The middle-class model of masculinity was the Anglo-Protestant businessman, father, sportsman, and public official, who embodied the ideals of masculine self-improvement. Participation in voluntary societies offered male camaraderie, provided charitable relief in times of distress, and served as a forum for making business and political connections.

Working men's involvement in voluntary associations, amateur sports, and mutual aid societies alongside middle-class men has been understood in the narratives of some historians as providing working men with a collective experience that encouraged the formation of a collective working-class consciousness.[13] I suggest, however, that male associational life in Petrolia operated in a two-sided cultural context, where workers could be pulled in contradictory directions. Fraternal orders encouraged cross-class mutualism and provided the ideals of morality and class hierarchy that allowed the Knights of Labor to organize in Petrolia in the 1880s, while at the same time fraternal hierarchies, either office holding or higher degrees, resulted in the formation of elite groups.

An investigation of the making of a literary culture in Petrolia beginning in the 1870s reveals that although middle-class men dominated the public civic space of nineteenth-century Petrolia, women were found in public and wielded a measure of power. Civic ceremonies in nineteenth-century North America also displayed the public in deeply gendered imagery. Parades were encoded as masculine, and as Mary Ryan and Cecilia Morgan explain, women were cast as the dutiful spouses of male elites or as demure symbols of civic virtues.[14]

Fancy dress balls and annual civic celebrations on Victoria Day and Dominion Day served to legitimate the position of the town's propertied middle classes in the public sphere. Civic celebrations provided their male middle-class organizers with an opportunity to build a connection between the oil industry as the town's primary source of livelihood and middle-class ideas of respectability and citizenship. Ultimately this civic boosterism confirmed their status as community leaders while simultaneously looking to the economic interests of the community in the oil industry.

The volatility of the oil industry meant that when the markets slumped local drillers were unemployed. In December 1873 the first crew of foreign drillers left Petrolia for the Dutch colony of Tjibodas, at the foot of the volcano Tjarema, south of Cheribon in West Java.[15] The drilling crew left in the midst of the depression that plagued the country for much of the decade and into the next. During this period local oil producers and refiners attempted to agree on prices. The cartels formed nearly always collapsed. In 1898 John D. Rockefeller's Standard Oil Company gained control of Imperial Oil and transferred the refinery to the port community of Sarnia on the St Clair River. This significant blow to the economy of Petrolia was exacerbated a decade later when the Enniskillen oil fields began to dry up. As chapter 5 illustrates, some drillers found work in other regional industries, but the opportunity of adventure for young men raised in rural and small-town Ontario, together with their highly specialized skills in oil drilling and the prospects of high wages, led many of them to foreign fields. In an era when only the wealthy, soldiers, and diplomats travelled abroad extensively, Enniskillen drillers travelled to the oil-rich fields of Java, Borneo, Sumatra, Persia, Galicia, Germany, India, Newfoundland, Russia, Italy, the West Indies, and the United States. The foreign drillers came to represent another dimension of Petrolia's identity as an oil town.

Drillers who learned their trade in the oil fields of Enniskillen Township journeyed to oil fields all over the world, spreading their knowledge of "the Canadian drilling system," often bringing along with them the components of the drilling rigs, which were manufactured locally at the Oil Well Supply Company. The local economy of Enniskillen Township benefitted from foreign drilling: from the manufacture of drilling equipment, which was shipped abroad; and from the wages of foreign drillers, which were sent home to support their families.

Travel writing gave Enniskillen's foreign drillers the opportunity to present themselves as authorities on travel to distant lands and other topics of general interest to their neighbours and family members back home. My approach differs from post-colonial literary scholars who have tended to rely on either published guidebooks or imaginative literature for their primary sources. I gathered the letters written by foreign drillers that were published in the two local newspapers, the *Petrolia Advertiser* and the *Petrolea Topic*, later merged as the *Petrolia Advertiser-Topic*, during the heyday of foreign drilling between 1874 and 1930. The letters composed by Enniskillen drillers working abroad can be distinguished from fiction by the expectation of their grounding in historical actuality. In addition, I located a few extant memoirs, a diary, and personal family letters and used them alongside the drillers' travel letters to identify and map the basic tropes of colonial discourses. Despite this quite conventional expectation, I used symbols, myths, metaphors, and other rhetorical strategies more often associated with fiction than with travel writing to establish connections between home, nation, and empire and to establish hierarchies of race, class, and gender power. The letters are a kind of conversation – albeit one-sided – where the relationship between colonizer and colonized was one of "co-presence, interaction, interlocking understandings and practices."

I am indebted to the writings of post-colonial scholars of the English-speaking world, many of them literary scholars, who have attempted to understand the history of imperial meaning-making. A reading of scholarly works focusing on travel writing and the relationship between travel and the creation of racial and gender hierarchies, allowed me to identify the colonizing gestures in language. Rather than reading colonial texts as expressions of traditionally Western ideals, post-colonial literary critics read these literary texts as evidence of the manner in which such ideals have served the historical process of colonization.[16] Post-colonial theorists have also argued for a re-examination of colonialism, beyond a simple analysis of machines of imperial power exercised from European metropolises, to incorporate the interactive social dimensions of colonization that took place in the colonies of exploitation. Mary Pratt foregrounds the interactive dimensions of colonial encounters with an emphasis on how subjects are constituted in and by their relations to one another in the "contact zone."[17] This approach provides a useful way for thinking about how Enniskillen's foreign drillers went about constructing the domestic subject of European imperialism in the colonies of oil resources exploitation.

Home and empire have traditionally been constituted in British studies of imperialism as separate and distinct spheres, with home being the source of "Britishness," progress, and civilization, and Empire being the other side of the world, the "dark continent," the as-yet undomesticated space of cultural backwardness. A re-thinking of British imperial history by Antoinette Burton, John Mackenzie, Laura Tabili, Anne McClintock, Catherine Hall, and others has demonstrated that the relationship between "home" and "Empire" was dialectical rather than dichotomous.[18] In considering how the connections between "home" and "away," and "nation" and "empire," were shaped in Enniskillen Township, I argue for an appreciation of the peculiarities of the Canadian drillers both as British subjects of a white-settler colony and as colonizers in the colonies of oil exploitation. In the twentieth century, the dialectic of "home" and "away" forged by the foreign drillers has been used to build the image of Petrolia as "Canada's Victorian Oil Town."

With the removal of the Imperial Oil Company's refining operations to Sarnia in 1898, Petrolia was no longer the centre of the Canadian oil industry. Throughout the early decades of the twentieth century industrial reorientation was attempted with varying degrees of success. The exodus of skilled male drillers to "foreign fields" continued during the early decades of the twentieth century. As chapter 6 explains various heritage initiatives, relying on a narrative of local history constructed around Petrolia's origins as an "oil boom town," attempted to re-assert community identity.

Three Old Boys' reunions organized in Petrolia in 1908, 1925, and 1946 expose the connections between memory, performance, and place. The first was organized in the midst of the industrial reorientation that occurred in the aftermath of the relocation of Imperial Oil to Sarnia and the exodus of more skilled oil drillers to foreign fields. In the aftermath of the World Wars, Old Boys' reunions were held in 1925 and 1946 to bring "home" war veterans and foreign drillers and their families who had located elsewhere. In line with the civic boosterism of the nineteenth century, the Old Boys' reunions were intended to affirm the moral and social values of the planners who organized them as those of the community. These heritage events fit nicely into the growing historiography on the spectacle-culture of pageants, parades, and carnivals documented so well in the North American context in the studies by Nelles, Rudin, Radforth, Davies, Ryan, Walden, and Gordon.[19] As these scholars argue, establishing and maintaining community identity and public memory was contested terrain. Building on this body of research, my analysis of the three Old Boys' reunions illustrates how these festivals served to order individual and public memory.

In his theories of collective memory developed in the 1920s, Maurice Halbwachs saw that individual memories remain collective; social groups provide the environment in which individuals develop their memories. Halbwachs also forged an important connection between collective memory and place, going so far as to suggest that every collective memory unfolds within a spatial framework. He writes, "we can understand how we recapture the past only by understanding how it is, in effect, preserved by our physical surroundings."[20] More recently cultural geographers have referred to the "power of place," or place memory, in stimulating collective memory. Ordinary landscapes – towns, neighbourhoods, streets, and buildings – nurture collective memory by encompassing shared time in the form of shared space. As Delores Hayden writes, "places trigger memories for insiders, who have shared a common past, and at the same time places often can represent shared pasts to outsiders who might be interested in knowing about them in the present."[21] People's attachments to places are material, social, and imaginative. Petrolia's Old Boys' celebrations were particularly effective in claiming the symbolic and material importance of place by identifying Petrolia as the "old home town." History and geography served as a sort of "memory theatre," and a direct aid for recalling the past.[22] The participants were established in public space as part of a campaign to reassert a community identity based on the history of Petrolia as an "oil town."

The heritage initiatives in Petrolia after 1970 differed from those staged earlier in the century. In the case of the three Old Boys' reunions connections were forged between spectacle and place and a civic memory was formed. Beginning in the 1970s a new economic and commercial base for the town was built around the commercialization of heritage as a leisure service. In the context of deindustrialization and the post-1970s phase of capital restructuring that reduced the West's monopoly on industrial productivity, many localities developed a new consumer-oriented, tourist-heritage economy. In Petrolia various government initiatives, local community volunteers, and the town council's strategy of place promotion influenced the commercialization of heritage for tourism. The ongoing economic reorientation of Petrolia around tourism has been based on the preservation of the historic Petrolia Discovery oil field as a living museum, the conservation of the downtown Victorian façade, and the restoration of Victoria Hall as a venue for summer theatre. The writings by museum and heritage critics, particularly David Lowenthal, Kevin Walsh, Brian Osborne, Bella Dicks, Mike Wallace, and David Harvey, and by social historians Raphael Samuel and Ian McKay, have been especially useful in shaping my analysis of post-1970 heritage initiatives in Petrolia, particularly the issues fuelling "the heritage debate."[23]

Critics of heritage, such as Kevin Walsh and David Harvey, argue that since World War II, museums and heritage have contributed to a pervasive distancing from many of the processes that affect our daily lives in modernity, and more recently post-modernity.[24] Uprooted and unsure of the future, North Americans find comfort in looking back. Critics of heritage point to the pretensions and banalities of heritage and its tainted entrepreneurial character. As I point out in chapter 6, the re-creation of Petrolia as "Canada's Victorian Oil Town" to some extent anaesthetizes the past. Buildings have been, and continue to be, turned into artificial townscapes, commingling and confusing past and present roles.

Proponents of heritage demonstrate its potential to offer a representation of local life that is thought-provoking and accessible, and to provide a public forum for the expression of local identities. Bella Dicks refers to the duality of cultural displays intended for a mass audience. "Heritage and ethnic displays," Dicks writes, "provide representations for tourists, consumers and visitors, but these representations also address whose culture is on display." Through display sites, Dicks suggests, "People can ensure their own traditions continue, ones which might otherwise be lost, by educating people about them." Heritage, therefore, can be seen as the mobilization of identity for what John Urry identified as the "tourist gaze" as well as the "self-gaze" of those who are exhibited.[25] Building on this literature, I consider how local, regional, provincial, and national alliances influenced the development of heritage projects in Petrolia. The creation of a public past in Petrolia has been a political as well as a cultural process, involving the survival of a place identity, which required hanging onto Petrolia's history as "Canada's Victorian Oil Town."

This book draws on archival sources housed in the national and provincial archives, including the survey diaries, field books, and maps contained in the records of the Geological Survey of Canada and the extensive documentation produced by the Crown Lands office. I have relied mainly on local newspaper sources, however. Local newspapers reported on the vagaries of the local oil industry; the activities of voluntary associations and other local groups; plans for the organization of parades, celebrations, and other spectacles; and the comings and goings of the foreign drillers. Newspaper sources were supplemented by documents from local voluntary associations and records pertaining to town festivals and heritage initiatives found in the regional archive in the Lambton County Library in Wyoming, Ontario. At the Oil Museum of Canada and at the Petrolia Discovery I found primary sources pertaining to oil drillers who worked abroad, including oral histories, photographs, diaries, and workers' contracts. Local citizens, most notably Mr Charles Oliver Fairbank III, gave me access to documents in their personal collections. These sources allowed me to analyse the cultural importance of place and to develop a potentially unique perspective on the literature on resource communities.

What follows is a chronological and thematic story of the colonization of one Ontario township and the making of a small town during the nineteenth century around the discovery of a single resource: oil. Unlike earlier studies of colonization in Canada, which focus on nation-building from the top down, this study teases out the connections between culture, colonization, and industrial capitalism at the local level, where Crown officials, settlers, speculators, and manufacturers interacted and where the processes of class and community formation occurred. This study draws on a vast array of literature from cultural and literary studies and from Victorian social and cultural history to argue that colonization and the formation of community identity were social and cultural processes, as well as an economic process, that took place in a dialectic between "colony" and "Empire," and "home" and "away."

CHAPTER ONE

"Oil Mania": Colonial Land Policy, Land Speculators, and Settlement in Enniskillen Township, 1830s–60s

Colonial governance was transformed during the early decades of the nineteenth century in the settler colony of Upper Canada. An elaborate framework for the administration of the colony was put into place, accompanied by a shift towards colony-centred knowledge-production capacities. The township system of organization played an important role in the vision of spatial order and identity in Upper Canada and surveyors evaluated the potential of these lands for settlement as part of the work of administrative knowledge formation. Colonial officials sought to create a terrain for capital accumulation through dispossession of native peoples, transformation of land tenure towards freehold, and registration of property. Nineteenth-century practices of geographical reconnaissance and land registry made Upper Canada amenable to European settlement and colonial administration. According to the "civilizing" process imagined by colonial administrators, the wilderness frontier would give way to bounded property comprising a landscape of family farms and compact little villages populated by loyal and industrious British subjects.

An act establishing registry offices in each county was passed in 1795 to record all transactions and agreements pertaining to property. In addition to records of sales of land and financing of mortgages, documents were created pertaining to the sale or lease of water or mineral rights. Land policies in Upper Canada did not affect every township equally.[1] Colonialism was never a uniform or coherent imposition, but rather a profoundly mediated set of relationships between local Crown land agents, settlers, squatters, and speculators. Most settlers dealt with the regulatory regime of the state at the local level. The local Crown land agent allocated land and managed the sale of lots. Crown land agents had considerable leeway and imperial regulations were interpreted from colonial vantage points.

The free-grant policy was aimed at preventing land speculation in Upper Canada, but as G.C. Patterson writes, "That bird of evil omen, the land speculator, was not long in appearing on the scene." Land speculators were marginalized by Colonial administrators.[2] Oil resource extraction uniquely shaped settlement and land speculation in Enniskillen Township, from the opening of the township to European settlement in 1835, through the "oil boom" of the 1860s, when speculation in oil lands was rampant. In this chapter, I examine how township lots as bounded property were alienated from the Crown during a period when Enniskillen Township was a rather unattractive site on the frontier of agricultural settlement. The latter part of the chapter explains how land speculation was carried out in Enniskillen Township at the height of the oil boom of the early 1860s. I suggest that during the early stages in the development of the oil industry the activities of land speculators were part-and-parcel of the normal workings of capitalism.

A number of assumptions have been implied in the historical literature about Upper Canada's land speculators, and the historical writing on Upper Canada includes some debate about how speculators were actually identified. David Gagan, A.G. Brunger, and John Clarke used a "scale of operation" approach.[3] These scholars defined land speculators as those who acquired large parcels of land, usually 400 or more acres, never intending to develop it themselves. Land was purchased cheaply with the expectation that the undeveloped land could be resold at a profit. Randy Widdis criticized the "scale of operation" approach and suggested that speculation could be defined only by the "motivations" of the participants. In a similar vein, William Shannon argues that the historical use of the term "speculator" in Canadian historiography ignores the elements of risk and quick turnover. Land speculation, moreover, did not always involve large parcels of land.[4] In Enniskillen Township land speculation occurred in a limited geographical area, on small parcels of land where substantial quantities of oil were thought to exist. Thus, any approach to the historical study of land speculation in Upper Canada must also take into account variations that might be attributed to local geographic, economic, and social conditions.

SURVEYING ENNISKILLEN TOWNSHIP

In the aftermath of the acquisition of First Nation's lands by treaty, the surveying and mapping of Enniskillen Township commenced. On 13 July 1832, Surveyor General Samuel Hurd, acting on an order from the Commissioner of Crown Lands, instructed Lewis Burwell to conduct a survey of Enniskillen Township.[5] Hurd instructed Burwell to keep a field book of his operation, "the plans of which the general report thereon must be returned with your accounts." Hurd's job was to examine the data produced by surveyors in the field. He was the archivist of the department, arranging and reporting on surveyors' plans, maps, field books, diaries, and accounts. The administration of colonization was built around local, colony-centred knowledge-production capacities that were aligned with the project of imperial aggrandizement. The archive created by the Surveyor General emerged as the focal point for the heterogeneous local knowledge of metropolis and empire.[6]

Surveyors created and defined the spatial work of empire. In his response to Hurd, Burwell indicated that he would proceed with the survey of Enniskillen Township immediately with the assistance of Eliakim Malcolm. Burwell had some experience as a surveyor and he came from a family of surveyors. He was the brother of Lt Col. Mahlon Burwell, who between 1814 and 1825 surveyed all or part of 24 townships north of the Thames River and in the counties of Kent and Essex. In addition, Mahlon Burwell carried out survey work on the Talbot Road. In 1819 Lewis Burwell assisted his brother, Mahlon, with the survey of London.[7] Following the passage of a new survey act in 1818, prospective land surveyors had to be examined on their knowledge of the theory and practice of surveying by either the Surveyor General or his deputy. Changes were also made in the way land surveyors worked. From 1819 township surveys were let out on contract, with payment either in the form of land or a combination of money and land, thereby formalizing payment in kind. "This new way of recompensing surveyors had a disastrous effect on the accuracy of the surveys," historian John Ladell remarks, "as both contractors and surveyors scrambled to get as much land surveyed as they could in any given season."[8]

Township surveys were intended to be consistent in quality and content regardless of the skills of individual surveyors. All township surveys were conducted from baselines and astronomical recordings made by predecessors. Attached to the letter from Surveyor General Hurd to Lewis Burwell ordering the Enniskillen Township survey was a plan "showing the relative situation of the said township." Burwell was given an "expected" map of the survey based on reports already on file with the Surveyor General's Office. This was part of the process through which the Surveyor General's Office managed the production of knowledge.

Each surveyor was ordered "to keep a journal and field book, inserting whatever is observable for its singularity and value, towards the public utility, as ... the quality of lands, and timber, etc."[9] The following features were recorded in the field notes at regular intervals: vegetation cover, type of soil, topography, streams, bogs and swamps, and the suitability of the land for agriculture. Surveyors were required to pay particular attention to conditions that would hinder settlement. Poor drainage was a major hazard and received considerable comment, as did the lack of a suitable watercourse for the construction of mills. Along with field notes, surveyors were required to keep diaries of the day-to-day operations in the field.[10] Both Burwell and Malcolm kept diaries while carrying out the Enniskillen survey. The surveyors did not write in their journals as they travelled through the bush; rather they waited until the end of the day to collect their notes and reflect. Thus the surveyors' diaries and field notes were subject to editing at all times. The entries in Burwell's diary reveal that he spent several days working on the returns of the survey. The field books and survey diaries were part of the textually mediated process through which British ideas of spatial order were imposed on the wilderness. The field diaries themselves had a certain broader cultural relevance because of their similarities with the hugely popular genres of travel and exploration writing.[11]

Malcolm's field diary is a topographical narrative structured chronologically and spatially. For each day of the survey, Malcolm noted the route taken and the vegetation observed. The surveyors used a standard unit of measurement, the chain of 100 links, equivalent to 66 feet. The job of the surveyor was to blaze markers and straight property lines that privileged the logic of the cadastral grid. Calibrations based on compass readings, astronomical recordings, and baselines made previously were used to define straight property lines. The surveyor was instructed to acknowledge "natural" obstacles, notably creeks, swamps, and rocks, but he was not to let them interfere with the task of conducting the survey.

The work of the surveyors was gruelling. They were beset by bad weather, problems in procuring provisions, a shortage of funds, and pressures to have the survey finished in short order. In his diary and final report to the Surveyor General, Burwell revealed that the cost of provisions for the Enniskillen survey were exceptionally high. Unable to procure salt pork locally, Burwell went to Port Dalhousie to purchase provisions. The teamster, who was well aware of the poor road conditions in the western part of the province, charged exorbitant rates to transport the goods.[12] Burwell and Malcolm completed the survey of Enniskillen Township in just 93 days, between 3 September and 4 December 1832, with the assistance of eight chain bearers and three axe men. One of the chainmen, Jacob Fish, a native, also worked as a guide. Paradoxically, while the directions given by the native guide were integral to the production of a representation of colonial space, his work ultimately led to the dispossession of native peoples from their lands and legalized imperial claims to the territory by redefining "land" as "property."

Every township survey, including the survey of Enniskillen Township, was wrapped in scientific ideology and implicated with the British imperial ideal of a settler colony based on an economy of forestry and agriculture. Surveyors "saw" the landscape in a particular way; one that was connected to broader European scientific and enlightenment ideals. They comprehended physical spaces as simple geometrical constructs. By reducing physical geography to its starkest geometric frame, order was imposed, thus beginning the process of "cultivating" the British North American wilderness.

The Enniskillen survey was carried out using the 2,400-acre sectional system introduced in 1829. Lines were run for every other side road allowance and the sidelines between every sixth lot, creating a section of 2,400 acres made up of twelve lots of 200 acres each. To reduce survey mileage only alternate concession lines were run. The surveyors commenced work in the southwest corner of Enniskillen where Malcolm placed a white oak post. Next they ran the line between Enniskillen and Moore Townships all the way to the rear of the township. The surveyors divided Enniskillen into 14 concessions with the lines running west and east and the side roads north and south, creating a total of 448 lots on 82,174 acres. The land surface of Enniskillen Township is broken by two tributaries, running in a southwest direction, roughly parallel to one another: the north branch of the Sydenham River, known as Bear Creek, and its tributary, Black Creek.

Inventory science laid a conceptual and practical foundation for the spatial ordering of British North America. Utilitarianism, historian Suzanne Zeller points out, was of crucial importance in justifying inventory science.[13] Derived from the writings of Erasmus Darwin, a relationship between botany and geography evolved during the nineteenth century, with practical utility as its essential organizing principle. Part of the nineteenth-century surveyor's job was to distinguish between "good" land for settlement and agriculture and "poor" or "desert" land, based on an evaluation of tree species and vegetation. The prevailing forest lore associated basswood, beech, cherry, hickory, oak, sugar maple, and walnut with "good land" potentially useful for timbering and for farming and settlement. These spaces were viewed as "natural resources." Alder, birch, black, ash, elm, soft maple, tamarack, and willow were associated with "wet" and "swampy" land, and hence unsuited for agriculture.[14] The first European settlers in Enniskillen Township occupied the prime lands along the northern township line on concession fourteen and along Bear Creek and Black Creek. Settlers were aware of the mapping of the township and deemed these lots good lands for agriculture as they were covered with forests of hard wood comprising beech, maple, elm, and hickory.

While it is presumed that nineteenth-century surveyors and mapmakers replicated the essential character of the objects they observed, Richard Edney points out that on occasion they had to reveal that they were not all-seeing and all-knowing.[15] In doing so they exposed the rhetoric of observation. Burwell's field notes reveal that much of Enniskillen Township to the east consisted of black ash swamp. They were forced to abandon the survey at two points on the eighth concession. Burwell wrote: "I found the water so deep and the swamp so very bad that it was impossible to run this line any further to the East. It is undoubtedly one continued swamp through to the Town line." They were forced to rely on the observations of those already settled in the region to determine the extent of the swamp and the timing of seasonal floods. The "Great Enniskillen Swamp" was largely a seasonal phenomenon: wet and impassable during the spring and autumn, and a hard, dry bed during the summer months.[16] Patches of land on a faint knoll broke the continuity of the swamp, and were classified by the surveyors as "good land" for agriculture.

Interestingly, Burwell did not note any deposits of surface oil. This is significant given that his mandate was to report any potentially "useful" resources. Either the surveyors did not pass the oil gum beds or perhaps, given that the usefulness of petroleum as an illuminating oil was not widely recognized, they did not see these deposits as having any commercial value.[17] In his final report to Hurd, Burwell indicated that the area was of little utilitarian value. Burwell wrote, "the Township is generally rather wet Land, and very heavy timbered, with a mixture of almost all kinds of hard timber, which is generally indicative of a heavy wet soil. The soil is generally clay ... The streams all take their rise from the swamp, and consequently [they] are not durable ... In the dry seasons they are quite dry, but in the wet seasons discharge great quantities of water ... From these circumstances ... they are not suitable for the erection of Mills."[18] The soils, consisting mostly of heavy Brookston clays, were naturally fertile if adequately drained. Extensive underdrainage was required before intensive agricultural development could occur. This required considerable outlay of capital for tiles and labour. Later, with the aid of municipal and provincial funding provided under the Ontario Drainage Acts of 1869 and 1873, and the Municipal Drainage Aid Act of 1873, a municipal drain was constructed spanning the entire length of every concession line in the township.[19]

Nineteenth-century practices of geographical reconnaissance made Enniskillen Township amenable to European settlement and colonial administration. The completion of the survey and mapping of Enniskillen Township in the autumn of 1832 imposed "official" lines on land, and in the process "land" was redefined as "property," with specific rights and obligations for those who settled in the township. Surveying and land registry were intertwined in the settler colony of Upper Canada, where issues pertaining to land ownership were of foremost concern.

FROM WILDERNESS LAND TO BOUNDED PROPERTY

In 1835 the Crown opened up Enniskillen Township for settlement, thereby continuing the practice of absentee patenting by official grantees that had been going on since the beginning of the century in the older settled districts of the province. A patent is a grant of privilege and property made by the government, and the grantor is always the Crown. In the early nineteenth century it was issued after the fulfillment of settlement duties and the payment of fees, and signified legal title to the land, the right to mineral and timber resources, and manhood suffrage. Enniskillen was a new township and within the first three years after land became available on the market fully one half of

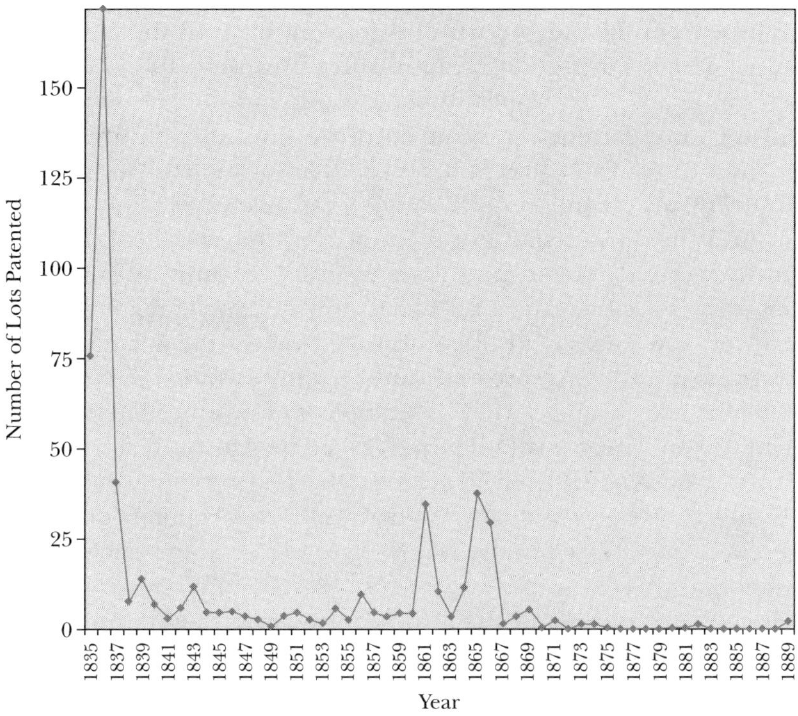

Figure 1.1 Rate of land alienation, 1835–1839. Patent Index, Enniskillen Township, Archives of Ontario

the lots were patented. Figure 1.1 illustrates the rate of land alienation from the Crown in Enniskillen Township from 1835 to 1889. During this period a total of 578 patents were issued. Those legislated by the Crown as having a right to land in Upper Canada, namely, military claimants, Loyalists, and the offspring of Loyalists, took advantage of the remission of settlement duties in 1835 and rushed to patent their claims. A total of 289 lots were patented between 1835 and 1837, and produced the following compilation: Loyalist grants, 174; military claimants, 81, heir and devisee grants, 10, sale of clergy reserves, 1; unknown, 23. Thus, 88.2 per cent of the patents issued by the Crown in Enniskillen Township went to official grantees. A second smaller wave of patenting occurred between 1861 and 1865, at the height of the oil boom. Another 15.9 per cent of the lots in Enniskillen were patented, most by assignment, where the locatee was not the first individual offered a specific parcel of land, and by the sale of clergy and Crown reserves and other sales.

Those from the oldest parts of the colony claimed the majority of official grants. Land grants in Enniskillen Township also went to satisfy the accumulated claims of old Loyalist and military settlements and as gifts or payments to members of the elite. Multiple grants were awarded to relatively few. Still, as J.K. Johnson noted, "Larger than normal grants of land were handed out as a matter of routine to anyone who could claim status or prominence."[20] Three prominent individuals received "larger than normal" grants of land in Enniskillen Township. Isaac Buchanan, a Scottish-born merchant, politician, pamphleteer, and staunch advocate of protection for Canadian industry, was granted 1,200 acres. Scottish author, John Galt, who was in charge of the Canada Company's field operations and spearheaded the settlement of a million acres of wilderness called the Huron Tract, received 1,100 acres. Samuel Hatt, a militia officer and businessman from Chambly, Quebec, was granted 800 acres.[21] A few commissioned militia officers who served in the War of 1812 were granted tracts of land in Enniskillen Township.

In most instances land changed hands quickly. Out of the 289 lots patented between 1835 and 1837 inclusive, 138 claims had passed to other absentee owners within five years. The colonial government's plan to settle the Western District with retired militia officers and Loyalists and their descendants was by and large unsuccessful in Enniskillen Township, as was the case elsewhere in Upper Canada. Land patents alone, however, do not distinguish actual settlers from absentee owners and land dealers. John Clarke questions the reliability of asserting any correlation between patenting and actual settlement because of changes in settlement requirements over time. He studied the spatial portrayal of patented and occupied land in Essex County using a cross-sectional analysis of patents and assessment rolls divided into three categories: (1) land identified as occupied in the assessment rolls or in the process of being patented with transactions occurring ahead of settlement; (2) patented and unoccupied land; and (3) unpatented and unoccupied lots.[22] A similar cross-sectional analysis was performed using the 1852 tax assessment roll for Enniskillen Township. This was the first assessment taken after Lambton became a provisional county.[23] The cross-sectional analysis of the patent index and the 1852 tax assessment reveals that of the 578 patents taken out between 1835 and 1889, 59.5 per cent of the lots were unoccupied in 1852. A considerable percentage of the patents, 21.1 per cent, were on lots that were not included in the 1852 assessment. Unpatented and unoccupied lots comprised 13.8 per cent of the lots assessed and only 5.5 per cent were occupied and in the process of being patented. As in Clarke's findings for Essex County, patenting clearly preceded settlement in Enniskillen Township. The use of land as capital by the majority of grantees, rather than for settlement, might, in part, account for sparse settlement. Changes in settlement regulations and difficulties in complying with these regulations, the swampy terrain, and inadequate drainage all limited the potential for settlement and agriculture. The high expense of establishing a family farm in Enniskillen Township deterred settlement. In addition, access to Enniskillen Township was limited to a rough track in the bush, known as the Nauvoo Road, between concessions eight and nine, with the line running between lots 17 and 18.

Unlike the settlers who paid their fees, completed their settlement duties, and secured a patent, squatters were categorized as bogus settlers because they had no legal title to the property. These semi-migrants stripped the best timber and took off a few good crops from natural clearings. On 6 September 1851, Peter Ward, who possessed a location ticket for the west half of lot 14, concession nine, wrote to the Crown Lands Office complaining about the state of the township. Ward stated that Enniskillen had only thirty-four actual settlers, but plenty of shanties used only by those who were plundering the Township of its most valuable timber and selling it in the American market. In his letter, Ward complained about the complacency of the settlers. He predicted that the day would come when the timber would be gone and suggested that "The Government would do well for the Country if they would give the land to actual settlers which would improve the Country."[24]

Among the first European settlers in Enniskillen were the Rouse, Oliver, Eveland, and Durance families. They probably arrived sometime before 1837. That year the militia officer came out from the Egremont Road and conscripted the male members from those four families who were capable of bearing arms to join the force guarding the St Clair River frontier.[25] Durance settled on lot 15, concession eight; Eveland on lot 14, concession ten; Oliver adjacent to them; and John Rouse and family at the site of the present-day Village of Oil Springs.[26]

Settlement in Upper Canada was not a levelling process, as inequalities in the initial distribution of land resources laid the basis for enduring socio-economic differences. Some historians have argued that colonial land granting policies actually prohibited the labouring-class immigrant from rising to the status of landowner.[27] The township papers in the Crown Lands Records indicate the difficulties experienced by settlers in Enniskillen Township and the strategies used by settlers in selecting, patenting, and developing their land. For "unofficial" grantees, who acquired land by paying fees and performing settlement duties, establishing a farm in Enniskillen Township was expensive and onerous. Some of the settlers were poor men whose struggles for propertied independence often turned into an elusive quest. In June 1843 the local Crown land agent returned the location tickets for two hundred acres on lot 29, concession two, stating that, "the party being a poor man, and the land in that Township being generally very low and wet," he had been unable to fulfill the settlement requirements.[28] Other settlers sought wage labour off the farm to pay their fees. In his petition for lot three, concession ten, Henry Clarke informed the Commissioner of Crown Lands that he was "a poor man with a large family, four boys and two girls with no means of support ... but working from place to place to earn a support for my family as we can get no support of [sic] this land."[29] According to the 1843 tax assessment for the Western District, only 270 acres were under cultivation in Enniskillen. In 1846 the total population of Enniskillen and Moore Townships combined was only 780.[30]

Enniskillen Township appeared as a separate district in the decennial census for the first time in 1861. The census manuscript provides one of the earliest extant descriptions of European settlement in Enniskillen Township. In the margins of the census form, enumerator John Smith remarked:

> A few persons settled in this Township as early as A.D. 1835, but for many years it was dreaded by settlers as a miserable swamp, and made very little progress. In the year 1854 the settlers had increased until there were fifty-four on the assessment roll, and the Township was erected into a Municipality under the amended Municipal Act ... Since that time the increase of the farming population has been steady and a good deal of improvement both in roads and land clearing, buildings, etc (also in schools) has been made.[31]

A total of 1,067 persons were counted in the 1861 census of Enniskillen Township: 670 males and 397 females. Only 15,283 acres or 18.6 per cent of the total area of Enniskillen Township had been taken up for farming, with 2,661 acres or 23.9 per cent actually under cultiva-

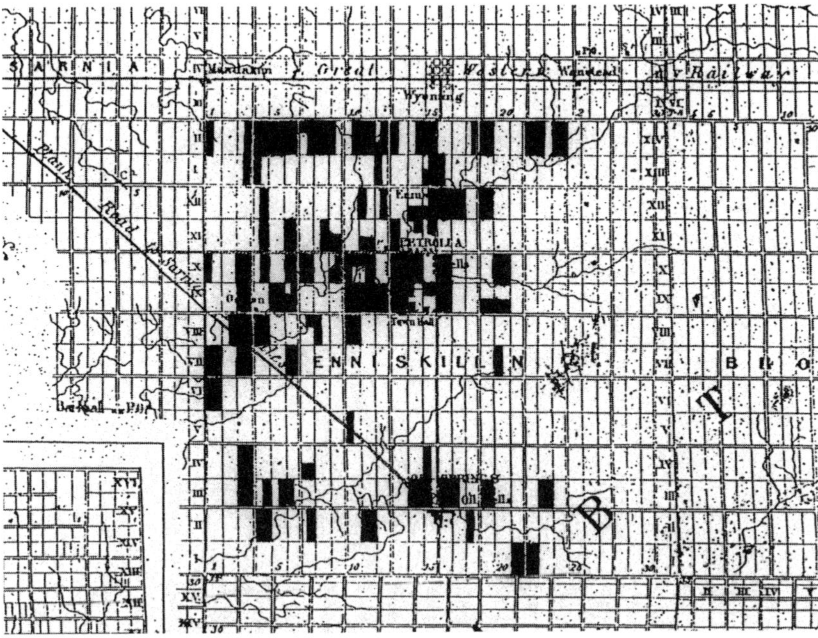

Figure 1.2 Farms in Enniskillen Township, 1861. *Belden's Illustrated Historical Atlas of the County of Lambton*, farms inserted

tion in 129 households. As Figure 1.2 reveals, the first areas settled were the well-drained, prime farmlands on the northern township line and along Bear Creek and Black Creek away from the swamp in the eastern section of the township, thus laying the basis for a "hierarchy of the soil." Successful family farms were built on these properties beginning in the 1850s.

The names and locations of 129 farmers were identified using the 1861 agricultural census manuscript, out of which 116 linkages were made with the Abstract Indexes to Deeds compiled by the County Land Registry Office. Analysis revealed that rather than any opposition between settlers and colonial administrators, the majority of settlers shared colonial administrators' masculine gender ideal for a settler-citizenry that embraced self-sufficient and respectable men of property. They completed their settlement duties, patented their property, and were considered "independent" of any other person's will by virtue of their property. Credit was extended to farmers by mortgages and, consistent with the findings by Gagan for Toronto Gore Township and by Clarke for Essex County, Enniskillen farmers entered hesitantly into formal mortgage agreements. When they did secure mortgages they paid them off as quickly as possible. Most of the capital generated from mortgaging came from within through private lending; institutional lending occurred infrequently during the 1850s and 1860s. Only 20 settlers, or 17.2 per cent, used primary mortgaging to capitalize the purchase of new or additional property. Secondary mortgaging was used more extensively in 47 households, or 40.5 per cent of Enniskillen farm households. As Gagan and Clarke suggest, secondary financing was probably used to underwrite capital improvement to the farm, acquire modern machinery, purchase livestock, or to see a farmer through a period of personal hardship.[32]

Settlers who were unable to pay their settlement fees or fully complete their settlement duties appealed to the Crown land agent for an extension. In these situations neighbours signed sworn affidavits before the Justice of the Peace attesting to any improvements made on the property and to the settler's compliance with ruling-class masculine gender ideals despite difficult financial circumstances. On 15 May 1855 George Wright and John Gaus signed a written oath before George B. Johnston, JP, that they were acquainted with the situation of lot 15, concession eleven, Enniskillen Township. William Anderson had taken possession of the property more than two years earlier, and had been an occupant ever since. Wright and Gaus swore that Anderson and his family had made improvements on the property: a log house had been constructed on the lot and five acres of land had already been cleared. The men also testified that William Anderson was of worthy character and a British subject. Although of "delicate pecuniary circumstances," he was viewed by his neighbours as sober and industrious, and was attempting to create a homestead for himself and his family.[33] Officially, however, the property had already been granted to another, a resident of Hamilton. The Crown Lands Office informed Anderson that he must abandon the property, but that he would be compensated for the improvements he had made on the land.[34] Settlers who complied with colonial ruling class ideals were generally treated with leniency by the colonial administration. Thus the pragmatics of settler colonialism were more complex than a simple opposition between colonial officials and settlers.

Although settlers came to Enniskillen Township with the intent of establishing family farms, John Smith, in conducting the 1861 census enumeration, commented that the "oil mania" brought the greatest influx of population to the Township. "It commenced in April last,"

Smith remarked, "and has tended greatly to increase the inhabitants, and to put a fictitious value on land, many parcels of land that a few years ago were thought hardly worth the taxes are now held at high prices." Smith referred to the prevalence of speculation in Enniskillen Township. He defined "oil speculator" as "the occupation of a number of well dressed persons who deal in oil claims; hunting up claims and getting leases."[35] The questions of how land speculation was accomplished in Enniskillen Township and of how to identify land speculators are integral to understanding the effects of colonial land-granting policies.

"OIL FEVER" AND LAND SPECULATION

Throughout the early part of the 1860s, the land market in Enniskillen Township was exceedingly volatile. The press regularly alluded to the activities of oil speculators, thereby pointing to the prevalence of the practice in the area. Newspaper correspondents typically presented speculators as dubious characters with a tendency to overindulge in alcohol. For instance, on 1 February 1861 the *Sarnia Observer* reported that although the temperature had been well below freezing for more than a week "oil fever" raged on. Its effects, according to the *Observer*, were "not like those of other fevers; the patient does not require to be confined to bed." On the contrary, "oil fever" patients might "camp out in some chinkless and unplastered log shanty, sleep in a corner on a bundle of weld hay or straw, wrapped in a horse-rug, and in clothing worn during the day, and which is, of course, well bespattered with mud, and redolent with the fumes of oil." To wind up this catalogue of "enjoyments," the *Observer* continued, they could "feast on unleavened cakes baked in the ashes, and pork 'spitted' in the smoke of a log fire ... the whole washed down by a decoction of strong tea, to which, mayhap, is occasionally added by way of seasoning, (if the patient be not a tee-totaller, – and we guess there ain't many of these among the "oil men") a dose of old Bourbon or Monogabela." The *Observer* cautioned that the benefit to the settlers of the township might not be as great as assumed at first glance and warned that, "The speculation will have a tendency to draw their attention from the work of digging the surface, and reaping the reward in the shape of fine farms, good crops, and superior stock, for the less certain employment of finding oil in the earth's bowels."[36]

Oil fever was accompanied by a dramatic increase in the price of property along the oil-rich banks of Black Creek and Bear Creek. In March 1861 the *Sarnia Observer* reported that land was changing hands rapidly in Enniskillen Township and sold at anywhere from $8 to $1,000 an acre, according to its supposed propinquity to the oleaginous deposit. That same month a special train was hired to allow land dealers from London to attend an auction of Crown and Clergy lands in Sarnia. Speculators were reportedly reaping rich profits by buying up tracts of land and selling them out in lots at enormous advances.[37]

The local populace tended to view oil speculators with derision. In February 1861 John Duff, from neighbouring Plympton Township, penned a letter to the editor of the *Observer* in which he disparaged the fact that nine-tenths of the oil lands were now in the hands of "foreigners – fully-blown Yankees; men who have been attracted hither by a thirst for mammon," and who cared little about the progress or prosperity of the country. Duff complained that European settlers born in Canada West and the Scottish immigrants among them had "allowed to grow up in their midst an extensive foreign monopoly of the recent discoveries in oil." He compared the speculation in oil land with the situation in the southern United States, where slave owners "look upon their colored victims merely as a source of profit, and as such only are the objects of concern."[38]

Duff's disdain for U.S. speculators was not shared by all of his contemporaries, however. His letter prompted a response from "A. Thrifty," also published in the *Sarnia Observer.* "A. Thrifty" pointed to the inconsistency of blaming "Yankee" speculators for exhibiting the promptness and energy of action that Canadians and Scotchmen were denounced for failing to show. He rejected Duff's argument and instead proposed that men of capital and possessing what he viewed as "rare and valuable qualities" should be encouraged to invest in the oil industry. "A. Thrifty" suggested that a "new" public opinion, along the lines of that taught in Belgium, Holland, and England, was needed. Men of capital were encouraged and protected in these countries, and whoever "develops the resources of a country, whatever his nation of origin, was a benefactor and not a robber."[39]

As the newspaper articles disclosed, and as historians Leo Johnson and John Clarke write, we can simply assume a priori that land speculation took place in Upper Canada, given the operation of laissez-faire capitalism.[40] A combination of homestead, male property right, and military bounty grants was intermeshed into colonial land policy in

Upper Canada. In the absence of money the state used land in lieu of cash as a reward for service. Many who received land in this way viewed it not as the means to establish an agricultural life, but as capital to be accumulated and spent as needed. Paradoxically, while colonial administrators marginalized land speculators, land-granting policies only succeeded in promoting speculation, particularly the provisions for Crown and Clergy reserves.

The scale-of-operations approach used by Gagan, Brunger, and by Clarke in his early research[41] neglects smaller scale speculation like that which occurred in Enniskillen township during the oil boom and ignores other transactions on the property, including subsequent speculative activity after the original grantee relinquished control of the land. Shannon and Widdis identified three categories of land speculators based on motivation: (1) land brokers who profited from a brokerage in land with a minimum of risk; (2) investors in land who bought land in the hope of capital gain by "sitting tight," holding onto property, and benefitting from the improvements made on neighbouring properties by other settlers; and (3) "quick flippers," who held their property for only a short period of time, usually less than five years, and who selected land with the greatest potential for high and immediate profit.[42]

In his recent book *Land, Power, and Economics on the Frontier of Upper Canada*, John Clarke identified land speculators using a two-pronged approach: (1) a simple summation of the acreage held by each individual; and (2) a reconstruction of the tenure status of each lot using the Abstract Index of Deeds for Essex County. Clarke concurs that scale is not the sole determinant of speculative activity, but he maintains that it is a useful determinant of power and economic influence. Equally important, Clarke suggests, were the number of transactions qualified by instruments through which property was obtained, namely, patent, bargain and sale, indenture, gift, mortgage, and sheriff's deed.[43] This approach combines number of transactions with length of time in the market and total acreage, thereby permitting a detailed examination of the broad zone along the speculator/investor continuum. The nineteenth-century use of the term speculator, therefore, included not only the absentee landowner of agricultural land but also the dealer in oil lots.

I used an approach similar to that used by Clarke, but retaining the categories of speculators identified by Shannon and Widdis, to study land speculation in Enniskillen Township. I examined every registered land transaction in Enniskillen Township from 1835 to 1869 using the Abstract Index of Deeds which were then linked with the Copy Books.[44] Individuals were classified either as land brokers, investors, or "quick flippers" according to the extent of their speculative activity measured by a summation of the acreage accumulated between 1835 and 1869, length of tenure, total number of transactions, and the location of the lots. A distinction was made between speculators taking advantage of colonial land policy who purchased, or who were granted their land by the Crown through the patent process, and those operating in the market economy. Individuals were identified as potential speculators if they held at least four hundred acres of land or engaged in a minimum of three transactions using an instrument of conveyance, either a bargain and sale, or a deed or an agreement, in addition to family exchanges of property. Although property did not actually change hands, a commercial transaction did occur when leases were signed for the development of oil properties. Hence leases and agreements were included when they were associated with oil resources.

This methodology indicated that motivation remains the key element in identifying speculators in Enniskillen Township. Individuals who purchased or were granted land in Enniskillen Township early on and had a long tenure had different motives from those who poured into the area at the height of the oil boom and held onto their investments for a short time. The scale of risk was different between 1835, when prices were low, and the 1860s when prices went up and declined again according to the availability of and market for oil. Also, speculation in oil lands in the early 1860s was largely, although not exclusively, confined to a particular space. The "territory" on which producing wells were sunk covered an area of about 2 1/2 square miles, and encompassed lots 15, 16, 17, 18, 19, and 20 on the first, second, and third concessions, where a thick deposit of oil had risen to the surface to form the "gum beds," and the "flats" on concession 10 around which the town of Petrolia was built.

This approach, as Table 1.1 indicates, yielded a total of 108 individuals engaged in land speculation in Enniskillen Township between 1835 and 1869. The forty-four "investors" were characterized by early entry into the land market and a long period of tenure. The majority of the investors, 84.1 per cent, secured property before the oil boom of the 1860s. Their activities were indicative of the "classical speculator," who obtained large but poorly developed acreage from the

Table 1.1
Land dealers in Enniskillen Township, 1835–69

Name of Speculator	*Background*	*Acres Assembled*	*#Trans-actions*	*#Patents*	*Acres Patent*	*Time Held (yrs)*	*Duration*
Land Brokers							
William Richardson	Oil Springs, Agent	626	32			4	1865–69
Investors							
Adolphus Mahon	London, Middlesex	1500	36	5	600	7	1954–61
Alex. Cameron	Toronto	1000	19	3	300	12	1857–69*
Alex. Manning (& Nathaniel Dick)	Toronto	1188.33	14			4	1865–69*
Alexander Dixon	Toronto, Esq., Sadler	500	14	3	500	20	1836–56*
Alexander Fraser (Hon.)	Charlottesburgh, Esq.	4200	30			16	1935–51
Alexander Graham	Oil Springs or Janesville, Wisconsin?	863.5	20			10	1855–65
Allan N. MacNab (Sir)	Hamilton, Esq.	2700	31			20	1835–55
Almond Buck	Hamilton, Gentleman; Coburg	600	7			10	1845–55
Andrew Elliott	Cornwall, Contractor, Esq.	5250	103	15	2900	8	1861–69*
Andrew Heron	Yorkville	1000	12			19	1847–66
Angus Carmichael	Wyoming	1756	55			9	1861–68
Angus P. McDonald	Hamilton, Contractor	800	5			9	1856–65
Anquilla B. Mitchell	Enniskillen, Yeoman	100	12			7	1861–69*
Asahel Haskin	Enniskillen, Yeoman	400	15			13	1849–62
David Thompson	Port Dover	400	5			26	1835–61
Frederick Wright	Belleville, Merchant	604	11			14	1851–65
George N. Ridley	Belleville, Surgeon	600	7	3	600	11	1846–57
George S. McPherson	Enniskillen, Yeoman	600	25	3	400	26	1839–65
George Stevenson	Sarnia	400	12			4	1865–69*
J.H. Fairbank	Oil Springs	404	15			4	1865–69*

Table 1.1 (continued)

James Henderson	Toronto	1950	27	5	950	24	1843–67
James Hervey	Montreal	800	22			28	1838–66
James Holmes	Montreal, Merchant	3000	14			27	1838–65
James M. Williams	Hamilton, Esq.	1800	41	1	200	12	1854–66
John B. Williams	Chatham, Esq	1000	13	3	400	10	1855–65
John Leys	Toronto, Engineer	681.25	17			4	1865–69
John Spiers	Montreal, Merchant	600	4			17	1838–55
John W. Sifton	Oil Springs	448.33	23			10	1857–67
Malcolm Cameron	Toronto (Hon.)	4700	123	2	400	27	1842–69
Michael Murphy	Enniskillen, Yeoman	663	36	1	100	10	1858–68
Oliver W. Chamberlin	Enniskillen	351	14			4	1865–9*
Patrick Barclay	Enniskillen, Yeoman	344.5	45			15	1853–68
Peter Carroll	Hamilton	1325	44			12	1854–66
Philip VanKoughnet	Cornwall	1600	13			20	1835–55
Robert Berrie	Inverness; London, England	600	13			10	1854–65
Robert King	Vaughn twp. Yeoman	850	13			27	1836–63
Robert Stanton	Toronto	400	7			16	1840–56
Thomas C. Street	Stamford, Wellend Co., Esq.	1600	31	1	100	18	1849–67
Thomas Forsyth	Sarnia	450	7			9	1856–65
Thomas Graham	Paris, France, Esq.	2800	44			27	1838–65
Thomas M. Jones	Toronto, Esq.	2800	38			20	1838–58
William Hutton	Belleville, Esq.	1100	7			14	1843–57
William Little	Yeoman, Enniskillen (London)	1190	22			11	1858–69
William Proudfoot	Toronto, Esquire	1100	21			29	1838–67
Zacheus Burnham	Hamilton, Gentleman	800	3			9	1838–47

Table 1.1 (continued)

"Quick Flippers"							
Aaron Choate	Hamilton, Gentleman	600	6			1	1837–38
Abel A. Adams	Erie, Penns, Enniskillen	295	15			5	1860–65
Abigal (Alizah) W. Farrar	Boston	317	5			5	1860–65
Abram Farewell	Oshawa	613	18			5	1864–69
Alexander Knapp	Chatham	400	6			3	1853–56
Alfred Boultbee	Newmarket	525	11	2	300	3	1866–69*
Alonzo Farrar	Boston	316.5	18			5	1860–65
Andrew J. Whipple	Oil Springs	608	24			4	1862–66
Benjamin Holmes	Montreal, Esq.	1300	12			1	1844
Charles A. Sadlier	Hamilton, Esq.	1000	14			3	1856–59
Charles A. Weller	Peterborough	600	10			4	1861–66
Charles K. Scholfield	Toronto	500	12			5	1853–59
Charles N. Tripp	Hamilton	1850	31	4	100	3	1853–56
Cosmore Bruce	Painsville, Ohio	60	4			1	1860
Donald McDonald (Hon.)	Toronto	400	40			4	1865–69
Edgar J. Jarvis	Toronto, Gentleman	600	8	4	552	5	1861–66
Edward H. Buche	Sarnia, Esq., Doctor	2243	11	3		3	1860–63
Ephraim W. Bond	Springfield, Mass.	1500	17			1	1865–66
Eugene S. Pike	Painsville, Ohio	200	20			5	1860–65
Frederick W. Watkins	Hamilton, Merchant	947	21			6	1860–66
George L. Thayer	Boston	220	20			5	1860–65
George Lowe Reid	Hamilton	1978	28			8	1861–69
Henry Benjamin	London	265	9			2	1865–67
Henry Tripp	Petersburg, Virginia, Gentleman	2798	7			5	1855–60
Hiriam Cook	Hamilton, Lumber Merchant	780	9			4	1856–60
Isaac Buchanan	Toronto, Merchant	1500	15	6	1200	1	1835–36
James Metcalfe	York	3884	90	9	1137	8	1861–69

Table 1.1 (continued)

James N. Scatcherd	Buffalo, Attorney	195	14			5	1960–65
Jesse H. Morley	Cleveland	1300	13			4	1861–65
John Brown	Hamilton, Merchant	400	8			4	1865–9
John C. Bullett	Philadelphia	600	12			1	1865–66
John Galt	Colborne, Esq.	1100	6	6	1100	1	1838
John Kemp	Enniskillen	133	21			4	1865–69
John L. Morris	Perth, Gentleman	600	5			4	1862–69*
John Macaulay	Kingston, Esq.	800	4			1	1836–37
Joseph Horrocks	Toronto, Gentleman	600	17	2	400	5	1861–66
Joseph Price	Hamilton	2209	37			3	1866–69
Joshua Adams	Sarnia, Attorney	920	20			3	1866–69
Leonard Stevenson	Enniskillen	125	17			2	1865–67
Melville Parker	Oil Springs	300	12			1	1866–67
Micajah L. Adam	Joliet, Ilinois	200	11			1	1865
Michael Shoemaker	Jackson, MI, USA	220	17			3	1868–69
Nathaniel Davies	York, Brewer	1100	20			5	1852–59
Nicol Kingsmill	Toronto, Barrister	380	10			4	1865–69
Oliver F. Farrar	Boston	283.5	12			5	1860–65
Peter Taylor	Moore Twp	217	33			4	1865–9*
Philip Ham	Belleville, Merchant	2400	23			3	1836–39
Robert Tindall	Toronto, Yeoman	877	18			7	1854–61
Robert Widdis	Oil Springs, Carpenter	167	12			1	1868–69
Stephen G. Lason	Pontiac, Michigan	141.6	9			1	1865–66
Stephen M. Jarvis	Toronto, Esq.	174	8			1	1865–66
Thomas A. Stayner	Quebec City, Esq.; Toronto	2000	3			1	1839
Thomas D. Ledyard	Toronto, Barrister	2099	47			5	1862–67
Thomas Fatzinger	Seneca Co., NY.	433.33	36			1	1865–66
Timothy B. Pardee	Sarnia	2256.66 6667	30	3	400	2	1864–66
Tristan Bickle	Hamilton	900	12			1	1865
Urbain Thibandeau	Quebec City, Gentleman	533.33	14			1	1863–64
Walter A. Dickson	Guelph, Gentleman	600	31	2	300	8	1861–69

Table 1.1 (continued)

William Darling	Montreal	300	9			1	1865
William E. Sanborn	Erie, Penn. Gentleman; Port Huron	325	27	1	300	2	1860–62
William Kelly	Erie, Penn.	1560.5	63			4	1860–64
William Kemp	Plympton, Yeoman	250	18			5	1860–65
William Richardson	Oil Springs, Agent	626	32			4	1865–9
William Stuart	Yeoman, Stormont	600	3			1	1838
Wm. Richardson	Enniskillen	560	35			4	1865–69
William T. Cooke	Chicago, Petrolia	100	12			1	1866

* Denotes individuals with land dealings after 1869

Source: Enniskillen Abstract Index, vol. A; Land Register, vol. C, Enniskillen Township; Copy Book Register B, Enniskillen Township, University of Windsor Archives

Crown. They did not show much discrimination as to where the property was located, and they were more likely to have obtained their property by patent than were the "quick flippers" who arrived in Enniskillen Township at the height of the oil boom in the early 1860s. The investors tended to sit on their property until the 1860s when they sold off part or all of their property in the mania surrounding the oil boom.

Among the investors in Enniskillen Township were several prominent merchants and politicians who amassed a considerable amount of property throughout Upper Canada, including Malcolm Cameron, Sir Allan Napier MacNab, and Alexander Fraser. For this group of absentee landowners, their property in Enniskillen Township was an obvious investment and a deliberate business strategy. As Douglas McCalla writes, early merchants in Upper Canada took on many roles: "they were at once partners and independent businessmen; importers and exporters, the latter on others' account as well as their own; wholesalers and retailers; transshippers at point of break-in-bulk; owners of mills and ships; dealers in land; creditors and financial intermediaries; and central figures in the province's external and internal communications."[45] These early investors acquired land as a kind of stock when the alternative was limited largely to Bank of Upper Canada shares.

The land dealings of Malcolm Cameron in Enniskillen Township were typical of the investor described by McCalla. The son of Presbyterian Scots, Cameron spent his early years in Lanark County where he entered into business as a general merchant. In 1835 he established a general store at Port Sarnia, and in 1837 he moved there himself. At Port Sarnia, Cameron set up lumber and flour mills and he built ships to transport goods along the Great Lakes. In 1847 he was a contractor in the building of the Great Western Railway. He also acquired 100 acres of what is now downtown Sarnia and subdivided the land into lots, some of which he sold in the 1840s and at a large auction in April 1857. In 1836 he ran as a moderate Reformer for Lanark County. In 1842 he was appointed Inspector of Revenue under the Baldwin-La Fontaine ministry, and in 1848 he became assistant Commissioner of Public Works under the second La Fontaine-Baldwin ministry.[46] Between 1842 and 1867 Cameron accumulated 4,700 acres scattered indiscriminately throughout Enniskillen Township. He engaged in a total of seventy-three transactions, sometimes mortgaging part of his holdings in Enniskillen Township, probably to raise capital to finance his other business dealings. He left his Enniskillen properties undeveloped or underdeveloped and operated as a manager holding onto his land for resale at a capital gain.

The speculative and other business activities of James Miller Williams in Enniskillen Township were also characteristic of the investor. In August 1858 he struck a flowing well on lot 16, concession two, after digging approximately fifteen metres into the clay loam. Williams was born in Camden, New Jersey, in 1818, and as a young man he was apprenticed to a local carriage maker. He moved to London, Canada West, in 1840, where he entered into partnership with Marcus Holmes to manufacture carriages. In the late 1840s he moved to Hamilton, where in partnership with Henry G. Cooper he formed the Hamilton Coach Factory.[47] On 3 February 1856, Williams, in partnership with Hamilton contractors Angus P. Macdonald and George H. Harris, Charles Anderson Sadlier, also from Hamilton, and Woodstock contractor John B. VanVoorhies, purchased 600 acres on lot 18, concession one, and lots 16 and 17, concession two, from the financially distressed Charles Nelson Tripp. On 1 December 1858 Williams acquired 600 acres on lot 18, concession one, and lots 16 and 17, concession two, part of the "gum beds," by indenture of bargain and sale from Charles Sadlier, thus beginning his twelve-year career as an investor and oil speculator in Enniskillen Township.

At first Williams refined oil on his property in Enniskillen Township. According to the *Sarnia Observer*, the illuminating oil produced at Williams's refinery was of a "superior quality." Its illuminating properties were reportedly so great that an ordinary-sized lamp could provide the light of six or eight candles. The illuminating oil sold for $1.50 a gallon.[48] Williams soon transferred his refining business to Hamilton, and in 1859 he was operating a refinery in that city under the name of J.M. Williams and Co. In 1860 Williams opened an office called the Canadian Oil Company at 18 MacNab St North, and a plant for refining oil was built at the foot of Wentworth Street, near the Great Western Railway line. Two years later, the company employed sixteen men, and turned out about 120 barrels of illuminating and machine oil per week. All of the crude was procured from the Enniskillen oil fields.[49] Discovery of a flowing well, plus the construction of a refinery, encouraged land speculation in Enniskillen Township and provided an impetus for the development of the local oil industry.

Williams, as Table 1.2 explains, accumulated 1,400 acres in Enniskillen Township between December 1858 and June 1861, most of it in the oil territory on concessions one and two. He also acquired a few lots scattered throughout Enniskillen Township, possibly with the intent of selling them for a profit should the oil deposits be found to cover a wider territory. Williams, in partnership with Messrs Anthony and Bush, subdivided his property on lots 16 and 18, concession one, and lots 16 and 17, concession two, and sold or leased these smaller lots to developers to raise capital.[50] Between 1854 and 1866, Williams was involved in twenty-nine transactions, exclusive of his extensive leasing arrangements on the sub-divided lots. Like many nineteenth-century businessmen, Williams had many diverse interests. He gradually passed control of the Canadian Oil Company to his son, Charles Joseph, and turned his attention to the manufacture of tin ware.

Andrew Elliott, a contractor from Cornwall, Upper Canada, was the largest landowner in Enniskillen Township during the early 1860s. He amassed 5,929 1/2 acres between 1861 and 1869. In the spring of 1861, Elliott was approached by a consortium of oil developers to provide the planks for a road between Oil Springs and the Great Western Railway line at Wyoming. An obstacle to the development of the local oil industry was the lack of a suitable road to transport crude and refined oil from Oil Springs to the Great Western Railway line at Wyoming. Between 1858 and 31 July 1861, only 5,529 barrels of oil were shipped by the Great Western Railway from Wyoming. During the winter months oil was more easily transported by sleigh, but during the summer months a team of oxen drawing a flat-bottomed "stone-boat" could haul only two barrels of oil along the mud road known locally as "the canal." Teams waded through mud up to their bellies, making the

Table 1.2 James M. Williams, land transactions

Con	*Lot*	*Instrument*	*Date*	*Grantor*	*Grantee*	*Quantity*	*Price*	
1	18	B & S	11-Oct-54	John Prince	James M. Williams	200	L 100	
1	18	B & S	1-Dec-58	Charles Sadlier	James M. Williams	200	L 1000	
2	16	B & S	1-Dec-58	Charles Sadlier	James M. Williams	200	L 1000	
2	17	B & S	1-Dec-58	Charles Sadlier	James M. Williams	all	L 1000	
1	18	B & S	19-Sep-59	Samuel W. Hawes	James M. Williams	5	$8,660	Rights to Drill
1	18	B & S	15-Feb-60	James M. Williams	John Fisher	2/3 of 5 ac	$2,000	
2	16	B & S	15-Feb-60	James M. Williams	John Fisher et al.	2/3 5 ac	$2,000	
2	16	B & S	14-Jan-61	James M. Williams	Canadian Oil Co.	part 5 ac	$2,000	
1	18	B & S	20-Apr-61	Angus P. McDonald	James M. Williams	200	L 100	
2	16	B & S	20-Apr-61	Angus P. McDonald	James M. Williams	200	L 100	
2	17	B & S	20-Apr-61	Angus P. McDonald	James M. Williams	200		
1	18	B & S	3-May-61	James M. Williams	William Anthony	1/2 200 ac	$20,000	
2	16	B & S	3-May-61	James M. Williams	William Anthony		$20,000	
2	17	B & S	3-May-61	James M. Williams	William Anthony	2/3 200		
1	16	Patent	5-Jun-61	Crown	James M. Williams	200		
1	18	B & S	14-Jun-61	James M. Williams	Canadian Oil Co.	part 5 ac	$2,000	
1	16	Plan	4-Sep-61	Samuel Peters, P.L.S.		200		
2	16	Plan	11-Sep-61	Oil Springs Plan No. 2				
2	17	Plan	11-Sep-61	Samuel Peters, P.L.S.				
1	18	Plan	26-Sep-61	Samuel Peters, P.L.S.				
2	16	Sispend	30-Nov-61	James Carroll	James M. Williams	all		
2	17	Sispend	30-Nov-61	James Carroll	James M. Williams	all		

Table 1.2 (continued)

2	16	Part. Deed	13-Dec-61	William Anthony	James M. Williams	all		
1	18	Agreement	14-Dec-61	William Anthony	James M. Williams	all		Partiton lands
2	17	B & S	14-Dec-61	James M. Williams	William Anthony			
1	16	Agreement	13-Jan-62	James M. Williams	Thomas H. McLean	2/3 200	$30	
2	16	Agreement	13-Jan-63	James M. Williams	Thomas H. McLean	all		
1	16	B & S	24-Feb-63	James M. Williams	Urban Thibandeau	200	$13,305	
2	16	B & S	24-Feb-63	James M. Williams	Urban Thibandeau	part	$13,305	
1	16	Mort	13-Mar-63	Urban Thibandeau	James M. Williams	200		
1	18	Mort	13-Mar-63	Urban Thibandeau	James M. Williams		$5,655	Disch
2	16	Mort	13-Mar-63	Urban Thibandeau	James M. Williams	part	$5,655	Disch
1	16	Assignment	2-May-63	James M. Williams	Calvin McQuestion	2/3 200		
2	16	Assignment	2-May-63	James M. Williams	Calvin McQuestion	all		
1	18	Agreement	13-Jan-65	James M. Williams	Thomas H. McLean	part	$30/ac	
1	18	B & S	24-Feb-65	James M. Williams	Urban Thibandeau	443.5	$13,305	
1	18	B & S	17-Mar-65	Henry B. Williams	James M. Williams	200	5 sch.	
7	5	Agreement	19-Dec-65	John Groom	James M. Williams	all 200	to sell	
10	23	Agreement	19-Dec-65	James Filman	James M. Williams	200	$2,400	
10	23	B & S	26-Mar-66	James Filman	James M. Williams	200		
11	8	B & S	3-Apr-66	David Thompson	James M. Williams	E1/2 100	$3,000	
10	23	B & S	16-Apr-66	James M. Williams	Robert Milron	W1/2 100	$2,600	
10	23	B & S	24-Apr-66	James M. Williams	George Roach	E1/2 100	$3,500	
11	8	B & S	31-Oct-66	James M. Williams	John Topping	E1/2 100	$5,500	
7	5	Release	9-Nov-66	James M. Williams	John Groom	all 200	$3,920	

Source: Land Dealers in Enniskillen Township, 1835–1869. Enniskillen Abstract Index, vol. A; Land Register, vol. C, Enniskillen Township; Copy Book Register B, Enniskillen Township, University of Windsor Archives

construction of a plank road absolutely necessary if the local oil industry were to succeed.[51]

The construction of the plank road was completed in the spring of 1862 in keeping with the agreement Elliott made with the group of oil developers. Elliott, however, was not paid the agreed-upon price of $20,000; he received only $13,000. In lieu of payment of the $7000 owed, he was given tolls for three years.[52] As a result, Elliott fell behind in the settlement payments on his land holdings in Enniskillen Township. In January 1865 he asked the Honourable George Brown to use his influence to secure an extension on the payments for his Enniskillen properties from Alexander Campbell, the Commissioner of Crown Lands.[53] Upon producing an affidavit signed by Reeve George McPherson, Elliott was granted an extension of six months. McPherson's affidavit provided details of the improvements made by Elliott to lots 14, 15, and 16 on concessions six and seven. The improvements included the construction of a sawmill worth $4,000, six dwelling houses, three stables, one barn, other outhouses, one blacksmith shop, approximately 20 acres cleared, an interest in the Wyoming and Enniskillen Plank Road running through the centre of the properties, and the sinking of two oil wells. Land speculation, as Elliott's activities reveal, did not merely imply "sitting on the land." It also involved settlement and improvement, thus revealing that the distinction between investor and speculator was not clear-cut, but rather a broad zone, as Clarke suggests.[54] Elliott was obviously a speculator who was also engaged as a producer in the development of his properties. As Table 1.3 further discloses, he mortgaged some of his lands as part of a private business of raising capital to invest in oil production and the sawmill, in addition to buying and selling land to finance his business endeavours.

Only one land broker could be confirmed from the analysis of the land registry records, although an examination of commercial and business directories suggests that there were more.[55] William Richardson operated as an oil agent for oil land dealers in the 1860s, managing leases and sales of oil lots much like a contemporary broker in stocks and bonds.

Most of the land speculators identified from the analysis of land registry records, 58.3 per cent (n=63), can be characterized as "quick flippers." They accumulated and disposed of oil lands quickly because of fluctuations in the demand for refined illuminating and lubricating oils on the market and uncertainty as to the extent of crude resources.

Table 1.3 Andrew Elliott, land transactions

Con	*Lot*	*Instru-ment*	*Date*	*Grantor*	*Grantee*	*Quantity*	*Price*
11	7	Ass't M	25-May-60	Malcolm Cameron	Andrew Elliott	E1/2 100 ac	$2,000
11	8	Ass't M	25-May-60	Malcolm Cameron	Andrew Elliott	W1/2 100 ac	$2,000
11	3	Ass't M	28-May-60	Malcolm Cameron	Andrew Elliott	E1/2 100 ac	Disch
2	23	B&S	1-Dec-60	Henrietta Sampson	Andrew Elliott	W1/2 100 ac	$600
2	23	Mort	1-Dec-60	Andrew Elliott	Henrietta Sampson	W1/2 100 ac	$450
9	7	Mort	Dec-60	Andrew Elliott	Allen Burton	W1/2 100 ac	$1,479
9	7	B&S	24-Dec-60	Allen Burton	Andrew Elliott	W1/2 100 ac	$1,500
2	5	B&S	2-Jan-61	Thomas Graham	Andrew Elliott	200	L 1050
2	5	Mort	2-Jan-61	Andrew Elliott	Thomas Graham	200	L 787.10.1
3	7	B&S	2-Jan-61	Thomas Graham	Andrew Elliott	all 200	L 1050
3	7	Mort	2-Jan-61	Andrew Elliott	Thomas Graham	all 200	L 187.10.0 other land
3	9	B&S	2-Jan-61	Thomas Graham	Andrew Elliott	200	L 1050
3	9	Mort	2-Jan-61	Andrew Elliott	Thomas Graham	200	L 787.10
8	3	B&S	31-Jan-61	Hugh Buchanan	Andrew Elliott	W1/2 100 ac	$300
2	12	B&S	6-Feb-61	John Dobbyn	Andrew Elliott	W1/2 100 ac	$920
3	11	Mort	22-Mar-61	Andrew Elliott	Thomas Graham	E1/2 100 ac	L 150
1	25	B&S	26-Mar-61	William Mattice	Andrew Elliott	all 200	$500
3	11	B&S	4-Apr-61	Thomas Graham	Andrew Elliott	E1/2 100 ac	
13	21	B&S	8-Apr-61	Fred'k Watkins	Andrew Elliott	E1/2 100 ac	$581
1	3	B&S	14-Apr-61	Edward G. Penny	Andrew Elliott	200	$2,272
1	3	Mort	30-Apr-61	Andrew Elliott	Edward Penny	200	$2,079
2	7	Mort	30-Apr-61	Andrew Elliott	Edward Penny	W pt 3 ac	Disch
11	3	Ass't M	4-Jun-61	Andrew Elliott	Malcolm Cameron	E1/2 100 ac	$2,000
11	7	Ass't M	4-Jun-61	Andrew Elliott	Malcolm Cameron	E1/2 100 ac	$2,000

Table 1.3 (continued)

11	8	Ass't M	4-Jun-61	Andrew Elliott	Malcolm Cameron	W1/2 100 ac	$2,000	
1	11	Patent	2-Aug-61	Crown	Andrew Elliott	E1/2 100 ac		
7	22	Patent	2-Aug-61	Crown	Andrew Elliott	200		
4	19	Patent	3-Aug-61	Crown	Andrew Elliott	200		
6	25	Patent	3-Aug-61	Crown	Andrew Elliott	200		
7	16	Patent	3-Aug-61	Crown	Andrew Elliott	200		
3	27	Final Order	27-Nov-61	Elizabeth Tait	Andrew Elliott	200		
2	13	B&S	27-Jan-62	Robert Sleed	Andrew Elliott	E1/2 & E1/2 of W1/2	$2,000	
1	3	Agree-ment	26-Mar-62	Andrew Elliott	Chas. Mackenzie	all		
1	25	Agree-ment	26-Mar-62	Andrew Elliott	Chas. Mackenzie	200		
3	27	Agree-ment	26-Mar-62	Andrew Elliott	Chas. Mackenzie	all 200	$15 p. ac.	
7	22	B&S	15-Apr-62	Andrew Elliott	Donald C. Thomson	E1/2 100 ac	$863.75	
3	15	B&S	3-Jun-62	Thomas J. Devore	Andrew Elliott	und 1/2 100	$2,350	
3	15	B&S	30-Jun-62	William Kelly	Andrew Elliott	und 1/2 100	$2,500	
2	20	Mort	21-Nov-62	William Kelly	Andrew Elliott	!/4 50 ac	$1,043.55	
2	20	Q.Claim	6-Aug-63	William Kelly	Andrew Elliott	all 200	$1	
2	20	Q.Claim	29-Jan-64	Mary M. Caldwell	Andrew Elliott	?	$1	
7	16	Mort	5-May-64	Andrew Elliott	George N. Carter	200	$3,658	other land
13	21	Mort	5-May-64	Andrew Elliott	George H. Carter	E1/2 100 ac	$3,658	
2	20	Mort	7-May-64	Andrew Elliott	Thomas J. Devore	und 1/4 of W1/2	$1,209.27	
7	22	Mort	7-May-64	Andrew Elliott	Thomas J. Devore	W1/2 100 ac	$1,209.27	
3	27	Mort	15-Sep-64	Andrew Elliott	Eliz & Marg't Tait	200	$1,510	
2	12	B&S	20-Mar-65	Andrew Elliott	Robert T. Elliott	W1/2 100 ac	$2,000	
1	11	B&S	30-Mar-65	Andrew Elliott	Robert T. Elliott	E1/2 100 ac	$7,000	

Table 1.3 (continued)

2	13	B&S	30-Mar-65	Andrew Elliott	Robert T. Elliott	E1/2 & E1/2 of W1/2	$7,000	
3	11	B&S	30-Mar-65	Andrew Elliott	Rob't T. Elliott	E1/2 100 ac	$7,000	
8	3	B&S	30-Mar-65	Andrew Elliott	S & Fl. Plank Rd.	2 ac pt W1/2	$1	
4	19	B&S	15-Apr-65	Andrew Elliott	Donald C. Thomson	E1/2 100 ac	$863.75	
1	4	Agree-ment	18-Apr-65	Wm. A. Rumsay	Andrew Elliott	200	$3,000	
3	15	Mort	23-May-65	Andrew Elliott	Richard Arnold	all	$436	Disch
3	15	B&S	29-May-65	Andrew Elliott	S & Fl. Plank Rd.	part	$1	
3	15	Mort	30-May-65	Andrew Elliott	James A. Wilkinson	200	$600	
2	20	Conveya nce	5-Jul-65	Andrew Elliott	Charles Heron	all 200	5 shillings	
3	15	B&S	9-Sep-65	Andrew Elliott	Robert T. Elliott	Lot 1 F 14	$250	
2	20	Q.Claim	5-Dec-65	Charles Heron	Andrew Elliott	Parts	$200	
2	23	B&S	8-Dec-65	Andrew Elliott	Thomas C.Chisholm	W1/2 100 ac		
11	12	Q.Claim	14-Dec-65	Mary Radenbury	Andrew Elliott	E1/2 100 ac	$200	
9	14	Deed	28-Jan-66	Andrew Elliott	Hector Cameron	17 1/2 ac of E1/2		
6	14	Patent	30-Jan-66	Crown	Andrew Elliott	200		
6	15	Patent	30-Jan-66	Crown	Andrew Elliott	200	$4,520	
6	16	Patent	30-Jan-66	Crown	Andrew Elliott	200		
7	14	Patent	30-Jan-66	Brown	Andrew Elliott	200	$4,520	
7	15	Patent	30-Jan-66	Crown	Andrew Elliott	200	$4,520	
1	3	Mort	3-Mar-66	Andrew Elliott	Nicol Kingsmill	200	$3,000	
1	23	Patent	9-Mar-66	Crown	Andrew Elliott	200	$600	
7	15	B&S	9-Mar-66	Andrew Elliott	A.J. Whipple	200	$8,000	
3	15	D of M	4-Apr-66	David Thompson	Andrew Elliott	200		
2	12	B&S	25-Apr-66	Richard Bell	Andrew Elliott	W1/2 100 ac	$4,000	

Table 1.3 (continued)

3	15	D of M	May-66	Richard Arnold	Andrew Elliott	200	
12	21	B&S	4-May-66	James Patterson	Andrew Elliott	W1/4 50	$2,500
11	12	Sispendus	11-May-66	Mary Radenbury	Andrew Elliott	E1/2 100 ac	
12	21	Mort	16-May-66	Andrew Elliott	James Patterson	W1/4 50	$2,500
11	12	B&S	30-May-66	Mary Radenbury	Andrew Elliott	E1/2 100 ac	$2,000
11	12	B&S	Aug-66	Rob't Tait Elliott	Andrew Elliott	E1/2 100 ac	$25,000
3	27	Patent	2-Aug-66	Crown	Andrew Elliott	200	
2	23	D of M	12-Aug-66	W.C. Samson	Andrew Elliott	W1/2 100 ac	Disch
9	9	B&S	13-Aug-66	Rob't Tait Elliott	Andrew Elliott	W1/2 100 ac	$20,000
9	9	Mort	5-Sep-66	Andrew Elliott	James Patterson	W1/2 100 ac	$2,680.76other land
12	21	Mort	5-Sep-66	Andrew Elliott	James Patterson	W1/4 50	$2,680.76
9	14	Sispen-dus	25-Sep-66	Elliott & McPherson	Livingstone & Chadwick	E1/2 100 ac	
9	14	Sispen-dus	11-Oct-66	Ananias Smith	Elliott & Ross	E1/2 100 ac	
5	6	Deed	31-Oct-66	Andrew Elliott	Joseph Elliott	200	$10
4	19	B&S	7-Nov-66	Andrew Elliott	Mart Elliott	W1/2 100 ac	$6,000
9	7	D of M	24-Nov-66	Allen Burton	Andrew Elliott	W1/2 100 ac	
1	3	Mort	28-Nov-66	Andrew Elliott	Nicol Kingsmill	200	$3,000
9	14	Dismis-sal	5-Dec-66	Court of Chancery	Smith vs. Elliott	E1/2 100 ac	
8	3	Mort	16-Jan-67	Andrew Elliott	Colonial Securities Co	W1/2 100 ac	$800
7	14	Mort	31-Jan-67	Andrew Elliott	McGarvey & Thompson	200	$825
9	14	Sispen-dus	22-Feb-67	Thos. Robinson	Andrew Elliott	E1/2 ac	
9	14	Deed	28-Mar-67	Thos. Livingstone	Andrew Elliott	E1/2 100 ac	5 sch.
9	14	Deed	29-Mar-67	Andrew Elliott	T.C. Livingstone	15 ac of E 1/2	$2,000
11	12	Sispen-dus	23-May-67	Mary Radenhurst	Andrew Elliott	E1/2 100 ac	

Table 1.3 (continued)

11	12	Dismis-sal	13-Jun-67	Mary Radenhurst	Andrew Elliott	E1/2 100 ac	
9	7	Mort	23-Jul-67	Andrew Elliott	James austin	W1/2 100 ac	$600
6	14	Mort	6-Nov-67	Andrew Elliott	James Patterson	W1/2 100 ac	5/-
6	25	Mort	14-Feb-68	Andrew Elliott	Nicol Kingsmill	E1/2 100 ac	$3,000
1	6	Agree-ment	13-Mar-68	Alex. Manning	Andrew Elliott	all	
9	14	Deed	14-Oct-69	Andrew Elliott	Walter M. Ross	60 ac of E1/2 & 1/2 ac of E1/2 & mort prem.	

Source: Land Dealers in Enniskillen Township, 1835–1869. Enniskillen Abstract Index, vol. A; Land Register, vol. C, Enniskillen Township; Copy Book Register B, Enniskillen Township, University of Windsor Archives

The tendency towards quick turnover of oil properties in Enniskillen Township started early in the development of the oil industry, beginning with the speculative activities of the Tripp brothers in the 1850s. Charles Tripp saw the potential for manufacturing asphalt from oil gum, and on 12 February 1853 he acquired lot 17, concession two, from Thomas Wait, an Illinois farmer, by an indenture of bargain and sale. Over the next several months, as Table 1.4 reveals, he acquired neighbouring lots on the "gum beds," including lot 18, concession one, lots 16 and 21, concession two, and lot 17, concession three – all by indentures of bargain and sale. The extent of the Enniskillen oil field was unknown at the time. Between 1853 and 1856 Charles Tripp amassed 1,450 acres in Enniskillen Township in twenty-three transactions; his brother Henry acquired 2,732 7/10 acres and carried out twelve transactions between 1855 and 1860. While they engaged in productive activity, mining for oil and manufacturing asphalt, the Tripp brothers also turned over some of their properties to raise capital.

The International Mining and Manufacturing Company was incorporated on 18 December 1854, with Charles Tripp as President. Hiram Cook, a wood merchant from Hamilton, John B. VanVoorhies, a wood merchant and contractor from Woodstock, and Henry Tripp, were chosen as directors. The company was capitalized at a value of $60,000 and each director was required to subscribe 250 shares with a par value of £1,250.[56] This venture in the manufacturing of asphalt

Table 1.4 Charles Nelson Tripp, land transactions

Con	*Lot*	*Instrument*	*Date*	*Grantor*	*Grantee*	*Quantity*	*Price*
2	17	B & S	12-Feb-53	Thomas Wait	Charles N. Tripp	E1/2 100 ac	L 250 Can
11	13	B & S	18-Apr-53	Luther Dunn	Charles N. Tripp	Pt 50 ac	L 250 Can
2	16	B & S	19-May-53	Augustus Jones	Charles N. Tripp	200 ac	L 6
2	16	B & S	2-Jun-53	Robert Graham	Charles N. Tripp	200 ac	L 112.10
2	16	Mort	2-Jun-53	Charles N. Tripp	Robert Garnham	200 ac	L 618.10
3	17	B & S	15-Oct-53	Alexander McNab	Charles N. Tripp	E1/2 100 ac	L 100
2	21	B & S	6-Dec-53	Charles Hendershot	Charles N. Tripp	W1/2 100 ac	L 50
7	13	B & S	17-Dec-53	William Wage	Charles N. Tripp	E1/2 100 ac	L 75
9	13	B & S	23-Dec-53	Thomas Graham	Charles N. Tripp	N 1/2 100	L 206.05
1	18	B & S	12-Jan-54	Joseph Raymond	Charles N. Tripp	200 ac	L 250 Can
3	17	B & S	21-Apr-54	John Rouse	Charles N. Tripp	W1/2 100 ac	L 125
3	17	Mort	21-Apr-54	Charles N. Tripp	John Rouse	W1/2 100 ac	L 50 Disch
4	11	Patent	21-Jun-54	Crown	Charles N. Tripp	E1/2 100 ac	
4	11	B & S	23-Aug-54	Charles N. Tripp	Andrew Stevens	E1/2 100 ac	L 58.10
2	17	B & S	3-Oct-54	James Beam	Charles N. Tripp	W1/2 100 ac	L 200
7	13	Deed	2-Feb-55	Charles N. Tripp	Henry Tripp	E1/2 100 ac	10 shillings
2	21	Deed	22-Feb-55	Charles N. Tripp	Henry Tripp	W1/2 100 ac	10 shillings
1	19	B & S	28-Jun-55	Gavin Nicholson	Charles N. Tripp	200 ac	L 250 Can
1	19	Mort	28-Jun-55	Charles N. Tripp	Oliver T. Maclem	200ac	L 150 Can
3	17	B & S	8-Sep-55	Charles N. Tripp	Henry Tripp	200 ac	L 400
10	13	Mort	14-Sep-55	Charles N. Tripp	A.P. McDonald	200	L 300
2	16	Mort	9-Oct-55	Charles N. Tripp	Edward McGivern	200ac	L 300
2	17	Mort	9-Oct-55	Charles N. Tripp	Edward McGivern	200 ac	L 300
1	18	B & S	22-Feb-56	Charles N. Tripp	Angus P. McDonald	200 ac	$2,000

Table 1.4 (continued)

2	17	B & S	22-Feb-56	Charles N. Tripp	Angus P. McDonald	200 ac	n.a.
1	18	B & S	22-Feb-56	Charles N. Tripp	Angus P. McDonald	200	$2,000
1	19	B & S	29-Feb-56	Charles N. Tripp	Theophilus Mack	S1/2 100	L 125
9	13	B & S	4-Mar-56	Charles N. Tripp	Hiram Cook	N 1/2 100	L 500
10	13	B & S	4-Mar-56	Charles N. Tripp	Hiram Cook	S 190	L 1500
11	13	B & S	4-Mar-56	Charles N. Tripp	Hiria	NE 40 of SE 1/4	L 1500
2	16	B & S	22-Oct-56	Charles N. Tripp	Angus P. McDonald	200ac	L 2000
1	18	Deed Poll*	22-Jun-57	Sheriff of Lambton	Charles Sadlier	200	L 170

* Tripp's land seized and sold to the highest bidder
Source: Land Dealers in Enniskillen Township, 1835–1869. Enniskillen Abstract Index, vol. A; Land Register, vol. C, Enniskillen Township; Copy Book Register B, Enniskillen Township, University of Windsor Archives

was not successful. Early in 1856 Charles Tripp began selling off his property in Enniskillen Township. In November he left the province, eventually making his way to the U.S. He abandoned his wife and in 1862 she asked the Chancery Court to replace Henry Tripp and Richard Martin as trustees of her properties in Enniskillen Township. Vice-Chancellor Spragge used the discretionary right of the court to appoint new trustees, but the court did not authorize her to take control of the properties herself.[57] In November 1866 the *Sarnia Observer* reported that Charles Tripp died in New Orleans on September 30.[58] Charles Tripp selected land in Enniskillen Township with the greatest potential for high and immediate profit, choosing sites that were oil rich in the hope of realizing a quick profit. In June 1860 Henry Tripp, who was also heavily in debt, gave up his claim to his properties in Enniskillen Township by "Quit Claim." He borrowed from Edward H. Buche, a Sarnia physician and oil speculator, but foreclosure on the properties was imminent.

A special correspondent hired by the Toronto *Globe* visited the Enniskillen oil region in September 1862. The reporter estimated, conservatively by his own account, that no fewer than 1,600 people had gathered together at this spot. Nearly every shanty contained its quantum of boarders. Americans, for the most part, although more Canadians had appeared recently. "There are not a few Californian miners, whose experience in gold digging materially assists them in their search after 'grease' as they call the oil," the reporter continued. "All sorts and all conditions of men" congregated at Oil Springs: doctors, tailors, colonels, and cobblers. "Many of them come to get employment and they are sure of it," the reporter noted, "others come with dollars in their pockets, to purchase land, and in a few days they have added others to the large number of wells already sunk or in the course of being sunk."[59]

A strategy of purchasing property and selling or leasing plots to oil developers was used by the "quick flippers." For instance, on 28 May 1861 George B. Cook and J.B. Bradley, both from Pennsylvania, leased twenty-five lots from William E. Sanborn, also from Pennsylvania, on the west half of lot 18, concession two, for a period of ninety-nine years. Under the terms of their agreement, Cook and Bradley promised, within a period of eight months, to commence sinking a well or wells on the lot and to work the same with "due diligence and skill" and, if a "good and successful well" was attained, "to pump and work the same with like due diligence and skill," yielding and paying to Sanborn one-third of the oil obtained from the premises, this to be delivered within twenty days. It was agreed that if Cook and Bradley failed to obtain a successful well after "fair trial," they would have the privilege of abandoning the premises. Also, if they failed to obtain a successful well after one year Sanborn would have the right to take full possession of the premises.[60] The number of lease agreements in Enniskillen Township during the 1860s is staggering. A perusal of land registry records suggests that the agreement between Sanborn and Cook and Bradley was fairly typical. Newspaper accounts suggest, however, that not many of the "wildcatters" who leased lands in Enniskillen Township found oil or became wealthy.[61]

Enniskillen farmers engaged in land speculation during the 1860s and sold or leased parts of their property to oil developers. The leasing arrangements between farmers and oil operators illustrate how the oil industry was integrated with agriculture, much like agriculture and forestry in the eastern parts of the province. Farmers with property near the oil fields leased portions of their holdings to oil producers for a sum of money and a royalty, usually one-third of the oil produced. On 28 May 1861 Sanborn leased 100 acres on the west half of lot 18, concession two, from Moses Wilton, a farmer. Under the terms of the agreement, Wilton agreed to let Sanborn subdivide the land into one-acre plots for oil exploration and the construction of two roads to connect with the main concession road at the north end of the lot. Provincial Land Surveyor E.R. Jones marked out a plan for the subdivision, and one-acre plots were leased to various oil producers.[62]

Unstable markets, excessive supply, and consumer preference for the "sweeter" smelling American illuminating oil over the pungent smell of sulphur residue that lingered in oil refined from Enniskillen crude, hindered the development of the local oil industry during the early 1860s, and made for a highly volatile land market. Unlike the classical speculators, who held onto their investment in property for a considerable length of time, "quick flippers" held their property for only a short time, usually less than five years, and tended to acquire less total acreage.

Land speculation in Enniskillen Township escalated briefly after the infamous "Shaw Well," located on lot 18, concession two, came in on 16 January 1862. The well reportedly yielded 15,000 barrels a day for several days, and 660 barrels a day after the flow was contained.[63] A month later, on 18 February 1862, the Bradley well, approximately two hundred yards from Shaw's well, came in. Soon afterwards Hugh Black struck oil on the east half of lot 17, concession one, approximately one mile south of the Shaw and Bradley wells.

The Shaw Well was the first of the "great producing wells," and ushered in the first real oil boom in Enniskillen Township, accompanied by a wave of land speculation. The speculation in oil lands was limited to an area of approximately 2 1/2 square miles: lots 15 through 20 inclusive, on the first, second, and third concessions, where thick deposits of oil had seeped to the surface to form the "gum beds," and the area along Bear Creek on concession ten around which the town of Petrolia was built in the early 1860s. Most of the oil producers were "shoe-string" operators, typical denizens of any nineteenth-century resource community. As local historian Edward Phelps writes, "most of them had nothing to lose, lived on luck and swiftly departed in search of greener fields when the boom slackened off."[64] Optimism was high in the spring and summer of 1862, as the quantity of oil in the underlying strata of the Black Creek region seemed limitless. Nevertheless, oil producers complained of dull markets and steadily declining prices.

The almost universal cessation of the flowing wells in the early months of 1863, beginning with the Shaw Well, created a panic among oil producers, and their attention shifted to the question of supply. In the spring of 1863 James Miller Williams began selling off his properties in Enniskillen Township. With the oil business virtually at a standstill, a group of oil producers decided to put down a "test well" on property owned by William Sanborn. Boring was carried out with vigour throughout the summer of 1863. The project was abandoned in October, when at a depth of 600 feet no oil was struck.

The first petroleum wells, also known as surface wells, were dug until oil filled the pit, usually at a depth of between 45 and 70 feet. The sides of the wells, about 8 feet wide and 12 feet long, were cribbed with logs put together inside the well using the same technique that was used to construct log houses. In her history of technology on the Ontario mining frontier Dianne Newell explains that when the surface wells ceased to flow, or became intermittent, oil producers began drilling into the rock.[65] At first rock wells were "kicked down" by manpower using spring poles. As oil producers were forced to bore deeper and further into the rock, capital expenditures for drilling escalated.

By February 1865 there were signs of renewed activity in the Enniskillen oil fields. Large joint stock companies, most of them financed by Americans with experience in the oil fields of Pennsylvania, were formed. They bought up large blocks of land from small, financially strained producers, who had cheap and primitive machinery. These large companies had the capital and the machinery to bore deep into the rock for oil. One of the largest, the Wyoming Rock Company, was formed in New York, and had among its trustees W.F. Havemeyer and George Opdyke, both former mayors of New York. Although fairly typical of business practices at the time, a large nominal capitalization of $1,000,000 was set for the company. In April 1865, the *Globe* revealed that the company had already spent $40,000 on the development of 275 acres in Enniskillen Township.[66] Prospectors from Michigan poured across the border.

Producers who had abandoned their leases only a couple of years earlier returned. Disputes over whether "due diligence" was exercised under the terms of these leases and questions over ownership of lots were settled in Chancery Court. On 26 April 1865 the Commissioner of Crown Lands issued a memorandum of instructions for "Lambton Oil Sales," stipulating that squatters had no rights under the Crown; nor did the original purchasers who were in actual occupation prior to 26 November 1864, but that they may be permitted to purchase the lot if improvements of a substantial character had been made, consisting of at least four acres cleared and roads, coupled with a dwelling house and bona fide occupation.[67] The colonial administration, however, was not successful in its attempts to curb speculation and squatting by wildcatters in the oil fields.

Osgoode Hall barrister J.D. Edgar published *A Manual for Oil Men and Dealers in Land* in 1866. In his introduction Edgar indicated that: "The unprecedented activity in land speculation, that has lately sprung up in the oil regions of Western Canada, suggested to the author the urgent necessity, at the present time, of a Manual which would give to oil men some light upon the transactions of every day." The term "speculator" in Edgar's narrative was intended as a term of derision. He writes, "There are always men who are willing to enter into agreements and speculations by which others can be bound, and who endeavour to escape from liability themselves if the result of the enterprise happen to be unprofitable." In a separate chapter Edgar reminded his readers of the qualifications for a "good title." "The seller must either show by clear evidence that he, or the persons through whom the land came into his hands, had it in possession or ownership for sixty years back." He pointed out that all instruments by which land in Upper Canada may be disposed of or affected are put on record in the registry office of the county where the lands lie, and that "it is always necessary to make a careful search in these offices."[68]

In 1866 a combination of occurrences resulted in the end of the first "oil boom." The Fenian threat posed a concern for oil speculators, many of whom were of U.S. origin. Afraid that war between British North America and the United States was imminent, the oilmen fled back across the border. Also, early wildcatters had drilled too many wells close together, and the natural gas pressures were bled off. Because the wells were left uncapped they could not re-pressurize. In June 1867, Robert McBride, an Oil Springs flour and feed merchant, amateur poet, and outspoken critic of land speculation and the policies of the Canada Land Company, penned *To Oil Springs Birds of Passage.*

You land speculators, with oil on the brain,
When will you revisit those regions again?
Like swallows in summer, you're still on the wing,
To catch other flies at the opening of spring.[69]

CHAPTER TWO

"Oil Smellers" and "Professors": Science, Colonization, and The Oil Boom in Mid-Nineteenth-Century Enniskillen Township

The Victorian era was characterized by enthusiasm about the powers and achievements of scientific inquiry. Historians of science have illuminated how Victorians of every class, in a variety of locations, defined knowledge, ordered nature, and practised science. Science and culture were inextricably linked in the eyes of the Victorians. Natural science reordered the ways in which people saw nature, self, and society. Ideas about how nature might be ordered suggested ways in which society might also be changed. Observation and investigation, rather than theory, was the basis of Victorian empiricism. In a sense nothing was "natural" about Victorian natural science since the scientific gaze entails an ontological imposition on the world.[1] Like the township surveyors, nineteenth-century European travel writers who visited the Western District drew on ideas from the earth sciences acquired at "home" in their efforts to make the new colony knowable to their readers. Travel writers also pointed out the utilitarian and commercial value of mineral oil for their audience back home.

In the early decades of the nineteenth century, geology emerged as the "most idolised of sciences." The underlying assumptions behind this new science were that the earth had evolved over an enormously long period of time and that its development resulted from the operation of ascertainable natural forces. Progressive forms of knowledge, political economy, geology, and nebular astronomy were at the heart of nineteenth-century controversies about the changes brought about by industrial-capitalist transformations and British imperial expansion. The principles of the natural sciences were used to define an idea of progress that had become associated with materialism and human perfectibility.[2] The sequence of dominant types, Victorian geologists postulated, had culminated in man, and no higher advance would ever take place. A popular culture of science emerged in voluntary societies formed to debate scientific theory. Public lectures and scientific demonstrations, popular magazines, and pamphlets provided insights into the discoveries of science.[3]

Colonization, capitalism, and science intersected in the development of the oil industry in Enniskillen Township during the mid-nineteenth century. The production of scientific work involved the intersection of different "social worlds," had a local dimension, and acquired its shape and meaning by virtue of the spatial, social, and cultural circumstances in which it was made and used. During the nineteenth century professional scientists cooperated with local artisans and practical men in the field.[4] Paul Lucier and Hugh Torrens have argued that we must dismantle the false dichotomy created by historians between gentleman-specialists and prospectors in the field, and attend to the relations between geology and industry, as well as to the importance of applied geology to colonial political economy. Victorian science was enmeshed in the "the warp and woof of commercialism, colonization, and capitalism."[5]

Geological surveys were productive ways and means for governments to encourage the exploitation of natural resources without themselves going into the business of resource extraction. The Geological Survey of Canada was founded in 1842. William Edmond Logan, the Montreal-born son of Scottish immigrants was appointed the first director of the survey. The fieldwork conducted by the geological survey failed to discover any extensive coal deposits in British North America. Thus, Logan was particularly interested in the petroleum deposits in Enniskillen Township, to ensure continued funding for the survey.

In Canadian historical writing, Morris Zaslow's official history of the Geological Survey of Canada, published in 1975, follows the example set by early practitioners in the field, who focused on the activities of the leading scientists and the institutional dimensions of the organization of surveys.[6] Suzanne Zeller examines the ways in which ideas of science informed the outlook of British North Americans during the era surrounding Confederation. Unlike the historians who suggested that scientific classification in the colonies served the interests of British imperialism, Zeller argues that, "the geological survey opened a way for rudimentary forms of nationalism to find an amplified expression." According the Zeller, "The political, social, and intellectual origins of the Geological Survey of Canada were closely tied to a growing concern with the country's destiny.[7] In Zeller's interpretation the geological inventory of Canada was not a reflection of the British gentleman-amateur tradition, but was influenced by the strong utilitarian reasoning that its director, William Logan, brought from Edinburgh. For Zeller science served the project of nation-building, which was linked to the utilitarian ideals of the Scottish enlightenment. William Eagan uses the concept of metropolis and hinterland put forth by J.M.S. Careless to contextualize Canadian history. He suggests that Logan created a distinctively Canadian institution and geology within the structures of the North Atlantic triangle and in a metropolitan context. "Borrowing people, institutions and scientific constructs from Great Britain and the United States," Eagan concludes, "Logan had begun in a position of subordination, as hinterlands must do, and successfully crafted an independent institution and geology as an equal participating partner within the triangle, naming and controlling Canadian rocks."[8]

In addition to probing class relationships in the development of the sciences, British historians have also attempted to explain the role of the natural sciences in the extension and maintenance of imperial power. The fundamental activities of surveying and mapping encoded and displayed relations of British imperial power. Territories were acquired by establishing scientific "facts," constructing a nomenclature, and by naming "things." Robert Stafford and James Secord uncovered the ways in which colonial data influenced important geological debates in Britain.[9] Their research demonstrates the importance of the intersection between geology and imperial expansion. These works are criticized, however, for their tendency to present the empire mainly as a resource for science, an object of study, and a source of data. As Bruce Hunt writes, "The empire, or more broadly the world beyond the European metropolis, appears here mainly in a passive role, as something on which scientists act and from which they take."[10] Informed by the writings of post-colonial scholars, Hunt suggests that, "The scientific work done in colonial countries must be examined in its own terms and in relation to the development by such countries of scientific traditions and institutions of their own." The history of the natural sciences reveals that the imperial-colonial connection was not simply a linear one-way process of diffusion from metropole to hinterland. As Frederick Cooper and Ann Laura Stoler write: "Europe's colonies were never empty spaces to be made over in Europe's image or fashioned on its interests; nor, indeed, were European states self-contained entities that at one point projected themselves overseas."[11] It is useful, therefore, to explore how difference was crafted in the many ways people in Upper Canada refashioned British and U.S. ideas to construct knowledge, order nature, and practise science. The analysis of the relationship between science and the oil industry in mid-nineteenth-century Enniskillen Township in this chapter focuses on the forms of power lodged in particular institutions and worked out locally in the colonial context. The conflicting visions of scientists (both professionals and gentlemen amateurs), practical men, oil speculators, and settlers illuminate the varied and complex ways in which knowledge about nature was ordered and economic and cultural power was deployed. I argue that geologists and practical men working in the Enniskillen oil fields provided contributions to the science of geology that were not fashioned solely in the interests of British imperialism.

EARLY EUROPEAN REFERENCES TO OIL SEEPAGES IN UPPER CANADA

Among the Aboriginal peoples of North America, petroleum oil was a sacred medicine.[12] Long before the arrival of Europeans, they dug below the surface for oil. As Lieutenant-Governor John Graves Simcoe and his wife journeyed from Newark (later renamed Niagara-on-the-Lake) to Detroit in February 1795, he noted in his journal that "the Indians discovered a spring of an oily nature which upon examination proved to be a kind of petroleum."[13] The discovery was significant enough to warrant a comment from Lady Simcoe, who wrote in her diary that "A spring of real petroleum was discovered on the march by the offensive smell."[14]

In the early part of the nineteenth century, travel writer John Howison referred to the oil springs on the Thames in his *Sketches of Upper Canada*. He wrote:

> At one of the houses where I stopped to feed my horse they showed me a specimen of mineral oil, that is found in considerable quantities upon the surface of the Thames. It flows from an aperture in the bank of the river, and three or

four pints can be skimmed from the water daily. It very much resembles petroleum, being of thick consistency and black colour, and having a strong penetrating odour. The people employ it medicinally; and I was told that its external application proved highly beneficial in cases of cramp, rheumatism and other complaints of a similar kind.[15]

In a similar vein, an English farmer, Joseph Pickering, wrote in his travel narrative, published in 1831, that at Moraviantown "a singular spring of oil issues out of the banks of the rivers near here, on the land belonging to the Indians." He described the oil as having "the consistency and colour of tar, with a peculiar smell." "It is generally supposed to be coal tar (petroleum) arising from a bed of coal said to run across the country," Pickering continued. "The oil is gathered from the surface of the water (by Indians and others) by blankets extended and lightly dropped on the surface." It was sold to merchants for from two shillings, three pence, to four shillings and six pence a quart and shipped to all parts of Upper Canada and the United States. The oil was used "as a cure for rheumatism, sprains, etc., and is sometimes taken internally, in small quantities for strengthening the tone of the stomach and for other complaints."[16]

Nineteenth-century European travel writers recognized the economic potential of petroleum as a treatment for various ailments as they mapped geographical territories and staked a claim to local resources. They also brought ideas about the earth sciences from their home culture, and imposed them on the new landscape in their efforts to establish order over the wilderness.[17] Pickering, for example, brought British amateur interests in natural science with him to the North American colony and he was apparently cognizant of the theory developed by European geologists in the eighteenth century about the common organic origins of coal and petroleum.[18]

GEOLOGY AND THE ENNISKILLEN "GUM BEDS": THE CONNECTIONS BETWEEN SCIENCE, COLONIZATION, AND COMMERCE

Petroleum was commonly referred to as "rock oil" or "coal oil," the latter name derived from the theory that petroleum and coal had a common organic source. Professional geologists employed by the Geological Survey of Canada tried to locate the stratigraphical position of the petroleum formations. The research conducted by the geological survey depended on the observations of practical oil producers in Enniskillen Township, who aided in this investigation in their rush to bore new wells. Ordering nature and establishing the utilitarian value of the mineral resources of Canada was central to the work of the Geological Survey and the project of colonization. Determining the factors controlling the accumulation of petroleum was also a major practical and theoretical problem for oil producers. Diverse groups were involved in scientific work in the Enniskillen oil fields. Both professional scientists and oil producers sought to establish the extent of the oil resources and to determine how they were formed. Knowledge has an important local dimension; it is shaped by the physical, social, and cultural circumstances in which it is made. Professional scientists and practical men working in the Enniskillen oil fields provided unique contributions to the science of geology that were not simply fashioned in the interests of British imperialism.

William Edmond Logan was born in Montreal on 20 April 1798, the son of prosperous Scottish immigrants. At the age of 16 he was sent, along with his older brother Hart, to high school in Edinburgh.[19] Logan attended the University of Edinburgh for only one year, where his mentor in chemistry, Dr Thomas Charles Hope, likely taught him some basic geological concepts.[20] He left the university and went to work at the London counting house of his uncle, Robert Hart Logan, in 1917. Logan's biographer, Bernard J. Harrington, remarked that it is not known to what extent Logan devoted himself to "amusements of a scientific nature" during his years in London.[21] In 1831 he moved to Swansea, Wales, where he was appointed manager in charge of accounts at the Forest Copper Works. He soon became concerned with all aspects of copper smelting, which involved locating and purchasing supplies of copper ore and coal. Logan acquired a practical knowledge of mining and metallurgy, and he prepared topographic maps of the geology of South Wales. His data included subsurface information obtained from miners. Logan constructed horizontal cross-sections upon which the underground occurrence of the coal seams was mapped. In 1835 he helped to organize the Swansea Philosophical and Literary Institute to encourage the study of natural history and the preservation of antiquities. That same year, the Geological Survey of Great Britain was initiated, and its first director, Sir Thomas de la Beche, adopted Logan's maps of South Wales. William Logan's employment in Swansea ended in 1838 with his uncle's death.

Up until this point, Logan's career typified what British historians of the earth sciences have described as that of "practical provincial men": amateur natural scientists who also worked in industry.[22] Like many of his contemporaries in industrializing Britain, Logan's work required some geological knowledge. These men cultivated geology as a science and ploughed the profits of their study back into their professional work. "Brilliant as these individuals often were in their professions," Roy Porter writes, "the effective scope of applying their geological wisdom practically was of course highly personal and local."[23] Logan's geological work in South Wales, nevertheless, prepared him well for the task of carrying out a geological survey of Canada.

In 1841 the Natural History Society of Montreal and the Literary and Historical Society of Quebec petitioned Parliament for a geological survey. A sum of £1,500 sterling was granted to defray expenses in conducting the survey. Logan was interested in the position of provincial geologist and he secured references from leading British geologists, including Henry de la Beche and Roderick Murchison. Logan's initial appointment was for two years. The mandate of the Geological Survey of Canada was topographical mapping and mineral reconnaissance: to furnish a full and scientific description of the country's rocks, soils, and minerals; to prepare maps, diagrams, and drawings; and to collect specimens to illustrate the occurrences. To assist him in his task, Logan obtained the services of Alexander Murray, a former lieutenant in the Royal Navy, who had immigrated to Upper Canada and purchased a farm near Woodstock shortly before the outbreak of the Rebellion of 1837.

Logan was acutely aware that the survival of the Geological Survey of Canada depended on his ability to illustrate the survey's relevance for the economic development of the colony. Geological science would facilitate and rationalize an inventory of resources which, in turn, would result in the progress of the colony. Consistent with the ideals of utilitarianism, this translated into a search for such "useful" minerals as coal and iron ore. There were rumours of coal deposits in the Gaspé Peninsula, and Logan began his work here. The following season Logan continued his survey of the region and in his report called attention to "two petroleum and two sulphurous springs in the neighbourhood of Gaspé Bay." Ultimately, he concluded that there were not sufficient quantities of coal in Gaspé to render mining a profitable enterprise. One of the earliest achievements of the Geological Survey, however, was to focus on the economic possibilities of the petroleum springs of southwestern Ontario, long known to the Native peoples and commented on by Lord and Lady Simcoe in their journals.[24]

Logan borrowed people and scientific constructs from Britain and the U.S. He retained a working relationship with the leaders of British geology into the 1850s. He was a close friend of Robert Murchison and supported him in his dispute with Adam Sedgwick over the boundary between the Cambrian and Silurian systems.[25] Logan also developed a close friendship with Professor James Hall of New York. U.S. geologists asserted their geological independence by developing a distinctive local nomenclature for their rocks and creating their own institutions, most notably the state geological surveys. Historians Zaslow and Eagan argue that Logan took the nomenclature and institutional structures developed in Britain and the U.S. and crafted an independent Canadian geological survey and a sophisticated Canadian geology. Logan and Murray divided Canada into three well-defined geological areas: an Eastern Division of folded Paleozoic rocks covering most of the Eastern Townships and the Gaspé peninsula; a Western Division of flat-lying Paleozoic rocks extending west from Montreal to the Detroit River, and split into two parts by a band of older gneisses known as the Frontenac Axis; and a Northern Division of Azoic rocks, the Canadian Shield. Logan created his own distinctively Canadian nomenclature for the formations comprising the Canadian Shield: the Laurentian and Huronian systems.[26] Canada's potential was conveyed to a wider international audience through progress reports, articles, lectures, and exhibits at international fairs. Logan eventually succeeded in building a general-purpose science and exploration agency, and he founded a museum in which he displayed specimens of economically important minerals. He also secured funding to hire a chemist, and founded a laboratory.

In December 1846 Thomas Sterry Hunt was appointed chemist and mineralogist on the Geological Survey of Canada. Hunt was descended from New England Puritans. At the age of twelve he was forced to go to work to support the family after his father died suddenly. He was self-educated in science until 1845 when he became a laboratory assistant to Benjamin Silliman Jr, a Professor of Chemistry at Yale College. Under Silliman's tutelage Hunt broadened his grasp of science, and with Silliman's recommendation he was given the job of analytical chemist at C.B. Adam's Geological Survey of Vermont. Hunt saw himself as primarily a chemist, and he approached geology from a chemical standpoint, beginning with a chemical reaction and from it deducing the necessary geological situation. Upon his arrival at the Geological Survey of Canada in early 1847 Hunt began studying the crystalline rocks of eastern Canada, sulphate and phosphate deposits, the composition of mineral waters, and the oil resources located in the southwestern part of Upper Canada.[27]

In the Report of the Geological Survey for 1849–50, Hunt referred to "the asphaltum or mineral pitch" found on lot nineteen of either the sixth or seventh concession of Enniskillen Township. He wrote that "attention was first called to it by His Excellency Earl Cathcart, who gave specimens of it to the Commission; since then Mr. Wood, the late member for Kent, has kindly sent a mass of more than one hundred pounds weight ... It is said to be spread over an area of several acres, and from the specimens received it is at least two feet in thickness."[28] Minerals and fossils were routinely sent to the scientists at the Geological Survey by prospectors in the field, practical workers, and gentlemen amateurs. The established coterie of professional geologists took these findings and located them within their own scientific constructions.[29] For instance, Hunt noted that the specimen was "about that of the variety known as *mineral caoutchouc.*" He suggested practical uses for the mineral pitch found in Enniskillen Township: "The consumption of this material in England and on the Continent for the construction of pavements, for paying the bottoms of vessels, and for the manufacture of illuminating gas, to which it is eminently adapted, is such that the existence of deposits of it in this country is a matter of considerable importance."[30] The specimen was of interest to Logan, especially since the Geological Survey had not discovered extensive coal deposits and continued funding for the survey was predicated upon establishing its use to the economic progress of the colony.

Alexander Murray reported to Logan in December 1850 that the black, bituminous, brittle shale observed in the bed of the Sydenham River at Zone Mills, on the township line between Zone and Dawn, was part of the Hamilton group that covered the entire western peninsula of Upper Canada. This observation refuted the prevailing theory that petroleum was derived from coal. The Corniferous limestone found under the drift clay, and categorized as part of the Hamilton formation, was described as being well below the location of the workable coal beds in the Lower-Sulurian rocks; the latter designation following the British system of classification. Murray suggested that, since bituminous springs were known to exist in the townships of Enniskillen,

Zone, and Mosa, "it would appear probable that a belt of the formation stretches across the point of the peninsula and extends from Kettle Point and the Rivière au Sable (south) to the Rondeau, and may possibly skirt the coast of Lake Erie for some distance down."[31] An extensive examination of the locality was planned for the following season.

The central focus of Victorian natural science was observation. On 31 May 1851 Murray conducted fieldwork in Enniskillen Township, where he measured the extent of the bituminous bed located on lot 16, concession two. Consistent with the practice among geologists of the time, Murray pencilled notes as he surveyed the field, and at the end of the day he penned over them in ink. Thus, a geologist could edit his journals and images, although Murray penned over his field notes of Enniskillen without making any changes. In his field notes Murray noted that the bituminous bed was "of but small extent." He wrote: "It occupies the ground close to the surface being covered only by about one or two miles." Trial holes were dug at various points in the bituminous bed to assess the depth of the deposit. Murray's field notebooks reveal that he relied to a considerable extent on the observations of local settlers and practical men in the field for information about the location and extent of the oil resources in the region. John Rouse, one of the first European settlers in the area, informed Murray that "the Black Bituminous slates are to seen on the North branch of the Bear Creek on Lot 16, con. 12 of Enniskillen in the bed of the stream." Another settler, Jacob Arnold, told Murray that he visited the "pitch bed" with Joseph Woods, where they found black pitch after digging 30 feet into the white clay. In his field notes Murray indicated that high streams from the spring floods prevented him from surveying this particular site and another bed of black pitch located in Zone Mills.[32]

In his official report of the season's fieldwork to William Logan, Murray suggested that the black shales of the Hamilton group were probably more bituminous than those of the Utica slate analysed by geologists working on the New York survey. Murray's fieldwork, therefore, contributed to European colonization by establishing scientific order over the region, although he drew comparisons using categories derived from the Geological Survey of New York. Following Hunt, Murray reported that this bed of bitumen located on a bed of white clay on lot 16, concession two, of Enniskillen Township, had the consistency of "*mineral caoutchouc*." "It would appear to have a thickness of two feet over about twenty feet square towards the south-west," Murray wrote, "and from which it gradually thins towards the edge in all directions, varying in some parts along a low ridge which it forms." He cautioned, however, that "its extent does not appear to be so great as we were at first led to understand." According to Murray, the field did not exceed half an acre, extending five chains in a north-east direction, with a breadth of less than half a chain. The economic viability of the bituminous bed had not yet been determined conclusively. Murray reported that he had observed bituminous oil on the surface of Black Creek in two places on lot 17, concession three. He stated that settlers in the area had informed him that oil had been discovered at other parts further downstream. In his report he stated that he was unable to comment on the amount of material that might be collected daily, as the current from the spring run-off swept away the oil as fast as it rose.[33]

The commercial potential of the oil fields was not fully recognized in the early 1850s. Brothers Charles and Henry Tripp probably heard about the Enniskillen gum beds from Alexander Murray. Henry Tripp had worked as a plate photographer in Woodstock in the late 1840s and Alexander Murray owned a farm in the area. Charles Tripp had worked as a foreman in a stove foundry. His knowledge of minerals and geology were acquired on the job and from his own reading and exploration. Cognizant of the economic gain to be made from manufacturing asphalt from the oil gum in Enniskillen Township, Charles Tripp acquired lot 17, concession two, on 12 February 1854.[34]

Prior to the incorporation of the International Mining and Manufacturing Company on 18 December 1854, Charles Tripp hired New York chemist Dr Thomas Antisell and Thomas McIlwraith, manager of the Hamilton Gas Company, to analyse a sample of asphalt from the Enniskillen gum beds. Antisell's report illuminates how science was used to serve the purposes of capitalism. Antisell wrote that he had distilled the sample at the temperature of boiling water, which yielded "a moderate amount of a liquid resembling Benzole." He observed that, "The application of a higher heat drove off additional volatile liquids, and a considerable quantity of paraffin." In his report Antisell indicated that it "was a very valuable variety of Bitumen, and applicable to all the purposes for which this substance is now in such demand." He stated that the asphalt was suitable for paints, adhesive products, waterproofing materials, and might be used as an illuminant if distilled.[35] From his distillation of the sample of asphalt Antisell pointed to the widespread commercial viability of the material from the Enniskillen gum beds.

The International Mining and Manufacturing Company was established as a mineral exploration company with Charles Nelson Tripp as President. The text of the petition for the formation of the International Mining and Manufacturing Company states:

> Your petitioners have been for the last three years, at great cost and expense, exploring various sections of this province for asphalt, lead, copper, silver, oil and salt springs, and the said Charles N. Tripp is owner in fee of two large asphalt beds in the western district, six oil and two salt springs in the said district, also one lead vein in the County of Prince Edward, lead in the township of Bedford, lead in Belmont, lead and copper in Otonabee, and has mining rights, leases and privileges in various portions of this Province.
>
> That your petitioners have associated themselves together for working and development of the said mines and minerals and for bringing the same into the market, whereby much benefit must result to the Canadian public.[36]

These were ambitious ideas, historian John S. Ewing writes, and it is difficult to believe that Tripp lacked any knowledge of chemistry and geology. According to Ewing, "the mere recital of powers of the company implies some little knowledge of these subjects."[37] Charles Tripp was a practical amateur who saw natural science and colonial industrial development as inter-related. The Tripp brothers business in asphalt manufacture in Enniskillen Township failed in 1857, however, largely owing to difficulties transporting the asphalt to urban markets, the lack of a suitable refining procedure, and insufficient capital.[38]

The use of natural science to promote the colonial oil industry continued after James Miller Williams struck oil in the summer of 1858 by digging into the clay to a depth of approximately fifteen metres. On 5 August 1858 the *Sarnia Observer* reported that,

> An important discovery has just been made in the Township of Enniskillen. A short time since, a party, in digging a well at the edge of a bed of Bitumen, struck upon a vein of oil, which combining with the earth, forms the Bitumen. The conjecture respecting this flow of almost pure oil is, that it has its source far in the bowels of the earth – that ages have been required to form, by its finding a vent, the present beds of Bitumen – and that the supply of fluid thus accidentally discovered will continue an almost inexhaustible source of

wealth, yielding, at the lowest calculation, and with no greater flow than at present, not less than one thousand dollars per day of clear profit.[39]

Whereas Charles Nelson Tripp was interested only in the manufacture of asphalt, Williams manufactured illuminating and lubricating oil.

At first Williams refined oil on his property in Enniskillen Township.[40] He soon transferred his refining business to Hamilton, and in 1859 he was operating a refinery under the name of J.M. Williams and Co. On 4 November 1860 the company was reorganized and incorporated as the Canadian Oil Company, with capital set at $42,000. In addition to Williams, four others were shareholders: three brothers, John, William P., and Nathaniel Fisher, and Isaac Jameson. Williams held the controlling interest in the company at $14,000. The plant was located at the foot of Wentworth Street, near the Great Western Railway line, at Coal Oil Inlet on Burlington Bay. In the summer of 1860 the illuminating oil produced by J.M. Williams & Co. sold for 70 cents per gallon, and machinery oil for 60 cents per gallon. Williams advertised his products in the Hamilton *Spectator*:

Those who have any knowledge of the coal oil business are well aware that not a barrel of any of these oils has been manufactured either in Canada or the United States without great difficulties having been experienced and large amounts having been expended in order to overcome the offensive odor common to all oils. The odor has prejudiced many against them, and has hitherto prevented us from bringing out oils before the public. But recent experiments have been attended with great success, and have resulted in our obtaining a process by means of which we can now entirely remove the disagreeable odor.[41]

The advertisement tells the story of several years of experimentation to remove the strong odour of sulphide found in the illuminating oils produced from Enniskillen crude.

Williams, who throughout his career always associated with expert partners and advisors, probably turned to the work of Dr Abraham Gesner, a physician who, like other Victorian-era gentlemen amateurs, developed an interest in natural history and geology. He published three books on the geology of the Atlantic colonies.[42] Gesner also came to be regarded as the Canadian authority on refining. At some point between 1840 and 1850, he focused his attention on refining illuminating oil. He perfected a process for the production of a lamp oil patented as "kerosene." He sold the process to the North American Kerosene Gas Light Company of New York, and he went to work for the company as their chemist and geologist. Gesner did reconnaissance work for his U.S. employers and he made more than one visit to the Enniskillen oil fields in the early 1860s on their behalf. As Gesner's work for American speculators suggests, capitalist motives shaped the development of nineteenth-century geology. By the early twentieth century an entirely new sub-field of geology – petroleum geology – had emerged as commercial interests provided financial resources for research.[43] Gesner also wrote an instructional manual entitled, *A Practical Treatise on Coal Petroleum.* In the manual he suggested that the offensive odour found in Enniskillen crude could be removed by adding nitric acid to the oil, in the proportion of one pint of acid to one hundred gallons of oil, and at the same time adding one gallon of chloride of lime.[44]

James Miller Williams was not alone in recognizing the commercial potential of the Enniskillen oil region. In compiling the 1861 census, enumerator John Smith remarked that "the oil mania" brought the greatest influx of population into the township. "The problem to be solved at present," Smith remarked, "is does this substance abound in the crevices of the rock generally or is it confined to those spots where it has showed itself spontaneously."[45] The prevailing notion among oil speculators, popularly known as "creekology," was that extensive deposits of oil could only be found near creek beds. The scientists at the Geological Survey were interested also in explaining the conditions for the accumulation of oil in Enniskillen Township.

Hunt presented his anticlinal theory as an explanation for the accumulation of oil at a lecture delivered to the Board of Arts in Montreal in February 1861, an account of which was published in the *Montreal Gazette.* "He supposes the oil to come from the Silurian and Devonian limestone at considerable depths below which are in many places very full of bitumen," the *Gazette* reported, "and he thinks we are warranted in expecting these oils in different places over a great portion of the western peninsula." The *Gazette* explained, "It required, however, a peculiar arrangement of the strata to allow the oil to accumulate and flow out, and this will only be met with, in the lecturers' opinion, along lines of folding and disturbance, which the geologists have shown to exist in various parts of that region."[46]

Hunt explained his anticlinal theory of oil accumulation in more detail in "Notes on the History of Petroleum or Rock Oil," published in the July 1861 issue of the *Canadian Naturalist*.[47] In the article, Hunt revealed that he formulated his theory on the basis of fieldwork he carried out in the Enniskillen oil fields in December 1860 and local producers' records of well output. He discovered that about 100 wells had been sunk by James Miller Williams and others in the southern part of Enniskillen Township along the borders of Black Creek, and also about ten miles farther north on Bear Creek. According to Hunt, only a small proportion of these wells furnished tangible quantities of oil, but already some 300,000 or 400,000 gallons had been obtained. He observed that the oil from the wells on Black Creek was more liquid and less dense than the oil from Kelly's wells on Bear Creek. Hunt writes, "It is said that wells recently sunk to a considerable depth in rock have yielded an oil still thinner, lighter colored and less dense, which is prized as being more profitable for refining."[48] From his fieldwork Hunt theorized that the productive wells were located "along the line of a low broad anticlinal axis which runs nearly east and west through the western peninsula of Canada, and brings to the surface in Enniskillen the shales and limestones of the Hamilton group, which are there covered with a few feet of clay." Gas and oil, following hydrostatic laws, accumulate at the highest points, or the domes, along anticlinal folds. Hunt explained,

> The oil doubtless rises from the Corniferous limestone, which as we have seen contains petroleum; this being lighter than the water which permeates at the same time the porous strata, rises to the higher portion of the formation, which is the crest of the anticlinal axis, where the petroleum of a considerable area accumulates and slowly finds its way to the surface through vertical fissures in the overlying Hamilton shales, giving rise to the oil springs of the region. The oil is met with at various depths; in some cases an abundant supply is obtained at forty feet, while near by it is only met with at three or four times that depth, and sometimes only in small quantities. Everything points to the existence of separate fissures communicating with a deep-seated source.[49]

Logan and Hunt concluded that the petroleum of the Enniskillen region occurred on the course of the great Cincinnati anticlinal running from the western extremity of Lake Ontario by Woodstock along the Thames River following the route of the Great Western Railway to Chatham, and then along Pigeon Bay, on Lake Erie. Hunt's theory and his accompanying description of structural conditions in the Enniskillen fields provided a guide for exploration.[50] The mapping of the Cincinnati anticlinal helped oil prospectors to locate other flowing wells in Enniskillen Township as the interests of science and capitalism became more tightly enmeshed.

THE CESSATION OF THE "FLOWING WELLS" AND POPULAR GEOLOGY IN THE ENNISKILLEN OIL FIELDS

Most of the early oil producers were small operators, typical of any nineteenth-century mineral resource community. The first petroleum wells, also known as surface wells, were dug until oil filled the pit, usually at a depth of between 45 and 70 feet, and involved little capital investment for tools. When the surface wells ceased to flow, or became intermittent, oil speculators began drilling into the rock. Dianne Newell writes that at first rock wells were "kicked down" by manpower using spring poles. This method, however, was soon superseded by steam-powered drilling rigs operated by specialized, itinerant teams of drillers. According to Newell, the tools and the method employed in percussion drilling were those that had been developed several decades earlier by the salt drillers of West Virginia.[51] Surface wells were dug and finished for about $3 per foot in the summer of 1861, and the cost of drilling into the rock was $2.50 per foot. The oil obtained from the surface wells was thicker and less pure than that from the rock wells. Flowing wells were reached at a level of about two hundred feet, whereas the flowing wells of Pennsylvania were over six hundred feet deep. Although Enniskillen crude was more pungent than Pennsylvania crude, it was also much cheaper to extract.[52]

A newspaper reporter from the Toronto *Globe* visited the Enniskillen oil fields in August 1861 and provided a detailed description of the wells. Williams & Co.'s well on Lot 18, Concession two had been dug to a depth of 46 feet, and then drilled a further 100 feet into the rock. The well reportedly averaged 60 barrels of oil per day and had been in operation two years. In a number of instances, however, speculators began to drill with insufficient capital and, having to go deeper than they expected, had to give up before oil was reached. One successful partnership, Blila & Co., who had leased land on lot 18, concession two, from William Sanborn, ran a well that allegedly produced 520 barrels in the first 23 ½ hours after oil was struck. The owners were unable to find enough barrels to contain the oil, and some 40,000 gallons overflowed into the creek. About 350 barrels were pumped into an unfinished neighbouring well in an attempt to prevent further waste.[53]

On 16 January 1862 the "Shaw Well," located on lot 18, concession two, came in, thus providing added impetus for the fledgling Enniskillen oil industry. The well's owner, Hugh Nixon Shaw, was an Irish immigrant who operated a general store in Cooksville. The narrative of Shaw's career, as reported in the daily press, was one of "rags-to-riches." Shaw began to drill for oil in Enniskillen Township sometime in late 1860 or early 1861. Shaw apparently "spent a good deal of time and money fruitlessly." He excavated 45 feet of soil, then struck rock and bored 100 feet. According to the London *Free Press*, his "means were exhausted; hope almost extinguished; credit gone; he was on the eve of utter despair," then his well came in allegedly on the last day he intended to work at it. The *Free Press* described the scene:

> At about ten o'clock on Friday morning at 203 ½ feet from the surface, the "oil was struck," and it came rushing up with a will – not giving time for the workmen to gather up their implements – filling the well in fifteen minutes, and shooting up a column of oil some 20 feet in the air. Everything obtainable was procured, but all that could be obtained, capable of holding oil, was insufficient for the wonderful supply, and hundreds of barrels of oil were flowing around the well, over the road into the creek. And so it continued day and night until Monday morning, when, by the insertion of gas pipes into the well and carrying the pipes into the air some 15 feet, the waste was stayed; but not until eighty or one hundred thousand gallons of oil were lost.

Shaw's well was the first "gusher" in the territory, and the first of the "great producing wells" in Enniskillen Township. The well reportedly yielded 1,500 barrels a day for several days, and 660 barrels a day after the flow was contained by a system of pipes.[54] A month later, on 18 February 1862, the Bradley well, approximately two hundred yards from Shaw's well, came in. Soon afterwards Hugh Black struck oil on the east half of lot 17, concession one, approximately a mile south of the Shaw and Bradley wells. Since there was no means for checking the flow of oil, thousands of gallons of oil spilled into the creeks. Large vats and tanks were constructed to contain the oil pumped from the ground. William Sanborn's tank, located on lot 16, concession three, was 20 feet square and 17 ½ feet deep. Sunk into stiff clay soil, and cribbed and puddled to a depth of three feet, Sanborn's tank held 1,250 barrels, or 5,000 gallons of oil. These tanks contained oil much more effectively than wooden vats, which tended to leak the oil, or else oil penetrated the wood.[55]

In addition to the problem of getting the oil to market over the treacherous "canal," prior to the completion of the plank road between Oil Springs and Wyoming in the spring of 1862, oil producers were confronted with the problem of excess supply and a limited market. In July 1861 Hunt conducted fieldwork in the Enniskillen oil fields. He remarked, "It is impossible to say what amount of oil these wells would furnish if wrought continuously, but the supply seems to be enormous." "Meanwhile," he remarked, "there is no market for the oil, and many thousands of barrels are stowed up in tanks and pits awaiting purchasers."[56] Enniskillen oil producers also paid high prices to ship crude and refined oil to market. In North America, Enniskillen oil competed with the high grade, less odorous Pennsylvania crude. While there was some success in opening up the British market in 1862, this proved only temporary as North American oil was considered more volatile than the paraffin oil manufactured in Britain. Oil producers complained of dull markets and steadily declining prices. A short-lived price-fixing cartel, the Canadian Oil Association of Enniskillen, was formed by some of the oil producers in the spring of 1862.[57] Optimism was high throughout the spring and summer, however, as the quantity of oil in the underlying strata of the Black Creek region seemed limitless.

Optimism soon turned to despair, however. With the almost universal cessation of the flowing wells, beginning with the Shaw Well, in the early months of 1863, the attention of oil producers shifted to the question of supply. The failure of the flowing wells created a panic among oil producers. During the latter part of January, the price of crude fluctuated between $1.00 and $3.00 per barrel. For oil producers, the cessation of the flowing wells meant that they had to resort to the more expensive process of pumping to keep up supply. A year earlier, when crude oil was plentiful and sold for 25 or 30 cents a gallon, refiners could not help but make a profit. Local refiners suspended operations with the intent of forcing down the price of crude oil and many wells were abandoned for want of money for deep drilling.[58] The price of crude oil hovered between $1.50 and $2.00 per barrel during the summer of 1863. The *Observer* reported that the producers had "the best of it," where the previous year refiners were reaping large profits. The situation would be temporary, the *Observer* predicted, and "the manufacturers will in the long run make more than the producers of the raw material, and the merchants more than either."[59]

The oil business was practically at a standstill. Local oil producers recommended that a "test well" be put down to determine the extent of the oil supply. The editors of the *Oil Springs Chronicle*, Edwin T. Solis and T.J. Hudson, lamented: "Theories on the subject are as various as the action of the human mind, and however learned may be the dissertations on the subject of the origin of petroleum, or however plausible may be the theories concerning its source, it must be confessed that science has failed to reveal what is most important of all – the permanency of supply." They offered their own explanation for the cessation of the flowing wells. "For our own part, we do not regard the stoppage of the flowing wells as any indication of an exhaustion of the supply, but merely as an evidence of the exhaustion of the power which forces the oil to the surface." The question could only be determined by experiment. They proposed that a well be sunk to a greater depth than had yet been penetrated, "to a depth of 2000 or more feet if necessary," where most oil producers believed a great reservoir of petroleum existed. The venture would be costly, but could be financed through the formation of a joint stock company. According to Solis and Hudson, the greatest hindrance to the development of the oil trade was the caution manifested by capitalists in investing their money. This reluctance to invest, however, was owing to uncertainty as to the quantity and continued supply of crude oil.[60] The *Sarnia Observer*, on its part, criticized the director of the Geological Survey, William Logan, for failing to express an opinion as to the probable extent of the Enniskillen oil deposits. The *Observer* remarked that "The subject ... is of sufficient importance to warrant the Government to order a Survey and Report upon it."[61] Thus, science was viewed as serving the interests of industrial capitalism.

As conditions in the local oil industry continued to deteriorate, oil producers criticized the professional scientists employed at the publicly funded Geological Survey for their inability to determine the supply of petroleum resources. In late March 1863 a group of gentlemen residing in Sarnia, and a few others interested in the oil business, met and formed a committee to make an application to Parliament for funds to aid in the drilling of a test well. The proposition apparently stemmed from a widely held belief that many of those engaged in the oil business had exhausted their means. They argued that "a considerable amount was annually expended in Geological surveys, for the purpose of developing the mineral resources of the country, and as the question of Oil production and supply was one on which Geologists could give no satisfactory explanation, it would be very proper for the Government to step in and do what private enterprise would hardly feel warranted in doing." The *Observer* remarked, however, that the request for funds might be attributed to the condition of the market rather than to any actual deficiency of supply in the wells.[62]

The following summer, local oil producers put down a test well on property owned by William Sanborn. Boring was carried out with vigour throughout the summer months. The project was abandoned in October when at depth of more than 200 feet oil had still not been found.[63] Had they been successful in obtaining a supply of oil commensurate with the extra expense of deep boring, it might have given impetus to the trade. The failure to strike oil resulted in most small producers abandoning their leases, and oil speculation in the area abated.

Two years later, in February 1865, there were signs of renewed activity in the Enniskillen oil fields. Large joint stock companies, most of them financed by Americans with experience in the oil fields of Pennsylvania, were formed. They purchased large blocks of land from small, financially strained producers, who had only cheap and primitive machinery. These large companies had the capital and the machinery to bore deep into the rock for oil. The U.S. companies were exempt from taxes on crude and refined oil manufactured in Canada. As H.V. Nelles argues in his important study of the politics of resource development in Ontario, this exemption was part and parcel of "the manufacturing condition." The resource developers in Ontario were enthusiastic continentalists during the nineteenth century, since the U.S. was their best market and they had the capital and experience to initiate the proper development of the province's resources. The slow progress of resource development created the permissive condition that eroded the power of the state.[64] Prospectors from Michigan poured across the border, most of them gentlemen "who lived without labor" and who remained in the area only a couple of days.[65] Producers who had abandoned their leases a couple of years earlier returned, and disputes over whether "due diligence" was exercised under the terms of these leases were settled in court.

As speculators returned to the Enniskillen oil fields, lectures intended for an amateur audience of practical oilmen, popular geological works, and maps of the oil fields for prospectors were published. Recent literature on the history of science has criticized the positivist diffusionist model of the popularization of science for its characterization of knowledge as cultivated by elites and then watered down to reach lower classes. According to this model the scientist becomes the ultimate authority. As Bernard Lightman writes, "Popular culture can actively produce its own indigenous science, or can transform the products of elite culture in the process of appropriating them, or can substantially affect the nature of elite science as the price of consuming the knowledge it is offered."[66] Professional geologists, practical men, and oil diviners or "wizards," all participated in the construction of scientific knowledge in the Enniskillen oil fields.

During the 1860s professional geologists from Canada and the U.S. not only interacted with practical men in the field as they conducted fieldwork they also delivered public lectures to gatherings of oil speculators eager to profit from their investment. Because they were unable to determine the extent of the oil supply in the Enniskillen field, these "professional lecturers" were the subject of criticism from local oilmen. A lecture delivered in Oil Springs in January 1866 by Alexander Winchell, head of the geological survey of Michigan and a professor at the University of Michigan fuelled a critique of professional geologists by "Loi Reduc," the *Sarnia Observer*'s correspondent in the field. Loi Reduc remarked that "It would be a good thing if Professional Oil Lecturers were pensioned by the capitalists, and sent into retirement ... There are not a dozen persons at the Springs who have given any attention to the higher branches of study; who know Geology, except by name; and to whom a lecture would be, and is, mere jargon."[67] Loi Reduc disparaged "[t]his miserable smattering of school geology" in his correspondence for the *Sarnia Observer*. He pointed out that professional geologists had not been able to offer a consistent explanation for the anomalies in the strata accompanying petroleum. "Oil is a fact; geology is, and it is not," he wrote.[68] Loi Reduc's remarks also suggest, however, that oil producers took the specialized language of geology and made it their own as they sought to locate new oil resources.

Beginning in the mid-1860s popular books appeared that referred specifically to the geology of the Enniskillen oil fields, or else discussed the oil fields in the context of a discussion of the mineral resources of the colony. In 1865 Henry White published a guide entitled, *Geology, Oil Fields, and Minerals of Canada West: How and Where to Find Them.*[69] White's long professional experience as a provincial land surveyor and his experience in mineral exploration in Canada qualified him for the task. The Toronto City directory for 1877 lists his occupation as geologist and provincial land surveyor. It is likely that White was well versed in the literature on geology.[70] In the introduction to the book White writes that his objective was not to provide "an elementary or text book for the use of colleges and schools," but rather to provide a comprehensive geology of Western Canada, "whereby a knowledge of its geology and mineral producing formations could be acquired, without a life of labour and luminous reading." An additional objective was to offer a new theory for the origins of petroleum, its probable future supply, and how and where to find it. White revealed that in researching the book he drew on his personal experience in the field, the progress reports prepared by the scientists at the Geological Survey, and University of Toronto professor Edward J. Chapman's popular book on the mineralogy of Canada.[71]

Geology, Oil Fields, and Minerals of Canada West was intended for a readership of practical oilmen and was representative of the new cheap nonfiction works of the nineteenth century that provided useful knowledge. White explains the study of geology in terms accessible to oil speculators who probably had a limited knowledge of geological classifications and stratigraphy. Consistent with other nineteenth century popular writings in geology, a progressive narrative provided the organizing principle for the work.[72] White wrote:

> By studying the forces now in action, the manner in which they act, and the effects which they are daily producing and the application of the knowledge which these phenomena teach, to the explanation of similar effects from the same producing cause in past formations of the Earth's crust, we shall be able to understand the changes of which these effects are nothing more than faithfully recorded memorials, and if we can ascertain the relative order of time in which these events have occurred we shall have a connected, though undated, chronological history of all the great changes that have heretofore transpired or taken place in the world's varied and progressive stages.

Geology, White pointed out, was of "the most essential service" to the miner as it directs him to useful and economic minerals.[73] In White's narrative geology was presented as serving the interests of colonial industrial and commercial development.

White provided an overview, in ascending order, of the various geological formations of Canada West, beginning with the lowest strata of Niagara limestone. In doing so, he adopted the Geological Survey of Canada's re-classification of formations already identified by U.S. geologists to explain the formation of bituminous limestone in Canada West. Hunt's anticlinal theory pertaining to the conditions controlling the accumulation of petroleum was described in considerable detail. A low main anticlinal axis runs longitudinally through the western section of the Province, from the west end of Lake Ontario, and subordinate anticlinals likely ran parallel to it. "Enniskillen is on the north side of this anticlinal axis, but its oil springs may be, and very probably are, on a subordinate one, parallel, divergent, or connected with it," White writes.[74] He offers practical advice for locating the crown of anticlinals where oil was believed to lie close to the surface. The instructions for locating the peaks of anticlinals were printed in italics to emphasize the importance of this information for oil prospectors. White explains: "This is an easy process when exposures occur, or facts noted, in deep borings. For if, from the same surface level, several borings are made in an oil-bearing neighborhood in various given directions, and the several depths of the same strata noted, it would be easy to determine the inclination of the underlying strata, and where the crowns of the anticlinals are located, and may be found, and where borings may reasonably be expected to be successful in the production of liquid petroleum." White advised would-be prospectors to acquire a sufficient knowledge of the geological structure of the country, rather than go blindly and indiscriminately to work.[75]

In *Geology, Oil Fields, and Minerals,* White addressed the question of supply, since a few of the deep oil wells of Enniskillen had become intermittent or ceased flowing altogether. He rejected theories that petroleum resources would soon be completely exhausted, and advanced his own interpretation based on "chemical agency." White theorized "That it is being constantly generated and produced by some peculiar but unknown chemical action, in self-forming fissures and caverns, amongst the fossiliferous and other carbonaceous rock of the formations in which it is found, in quantities proportioned to the relative chemico-carbon-producing materials contained in the strata thus being acted upon … and that this chemical or formative process is constantly going on, supplying these cavities from whence it is obtained."[76] He noted that the science of chemistry was still in its infancy, and that he could not within the scope of his work give a detailed scientific dissertation to back up his theory. In putting forth his theory of chemical agency as an explanation for the formation of oil, White suggested that there was a relationship between geology and the other sciences, such as chemistry, mechanics, botany, zoology,

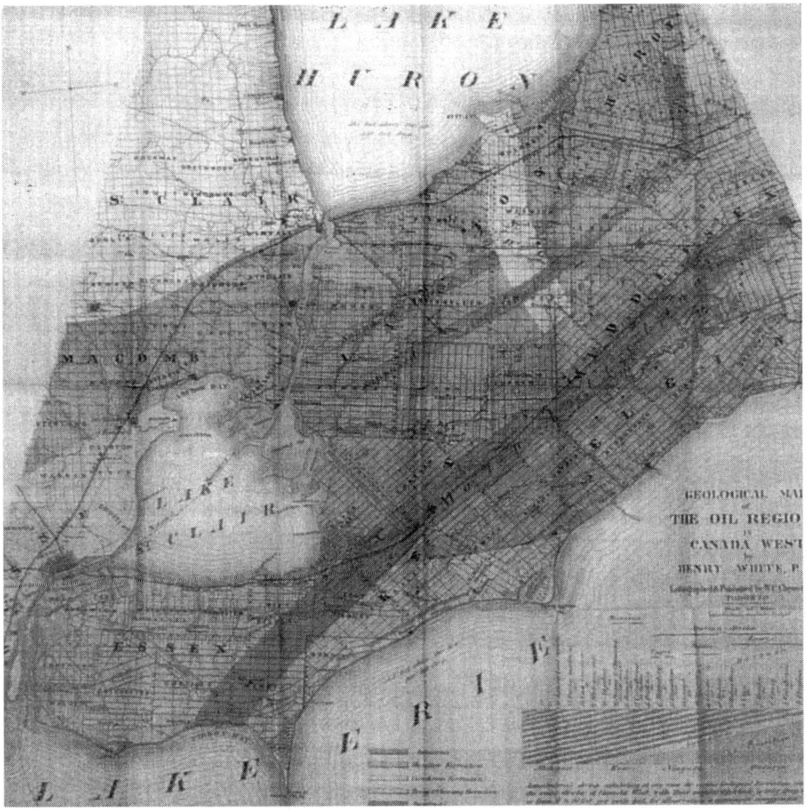

Figure 2.1 Henry White's geological map of the oil regions of Canada West. Henry White, *Geology, Oil Fields, and Minerals of Canada West: How and Where to Find Them*

comparative anatomy and mineralogy, and that these relations were those of mutual dependence.

White included a map of the oil regions of Western Canada to assist oil prospectors in locating areas where oil was likely to be found in substantial quantities. Visual communication was important to the development of geology in the nineteenth century, and it embodied a complex set of tacit rules and conventions. A geological map, Martin Rudwick argues, is an attempt to depict on a two-dimensional surface what cannot in reality be seen at all except in isolated exposure; it is "a highly complex, abstract and formalized kind of representation." It is also a subjective interpretation based on multitudes of observations, and the theories held when the map was made.[77] White's geological map of Canada West was prepared on a scale of five miles to the inch, with each lot and concession clearly delineated (see Figure 2.1). Following the conventions of nineteenth-century geological mapping, a series of colour washes were used to denote in broad outline the rock-types that characterized the various regions of Canada West. The colours were an important iconographical device because they represent an extrapolation from the evidence. As Rudwick explains, "the colours denote an implicit belief that the relevant rock-type would be found in the intervening areas, under the concealing cover of soil and vegetation." They also indicate implicitly the underground extrapolation into a three-dimensional structure.[78] A related iconographical structure was the provision of a key to the colours. The boxes in the key were placed closely together in a vertical column corresponding to the strata described in the text. Since the map was a thematic representation, the anticlinals where oil was believed to be located in substantial quantities were coloured boldly in red for emphasis. A detailed geological section was included in the lower right-hand corner of the map to provide an indication of the underlying geological structure of the area, although White indicated that the representation of the angular dips in the under-strata was "necessarily exaggerated."

White's book was intended as a practical field guide for oil speculators who had little or only a basic knowledge of geology. Alexander Winchell's objectives in *Sketches of Creation*, published in 1873, were different. Winchell's intentions are revealed in the book's sub-title, specifically, to provide "A Popular View of Some of the Grand Conclusions of the Sciences in Reference to the History of Matter and of Life." *Sketches of Creation* was aimed at two audiences: those who do not have specialized knowledge of science and students with a specialized knowledge of science who seek a broader, general perspective.[79]

Winchell was born on 31 December 1824, in the town of North East, Dutchess County, New York. He graduated from Wesleyan University in 1847 and embarked upon a remarkably diversified career of teaching, lecturing, and writing. Over the next several years he held a succession of teaching positions throughout the eastern and southern United States. Meanwhile he made extensive natural history collections and began what was to be a prolific output of scientific publications. In 1853 Winchell took up the chair of physics and civil engineering at the University of Michigan, and two years later he was appointed chair of geology, zoology, and botany, a position he was to hold until 1873. While teaching at the University of Michigan, Winchell wrote profusely, delivered numerous public lectures, and held the position of state geologist between 1859 and 1861, and again from 1869 to 1871. During the 1860s he visited the oil fields of Enniskillen Township and delivered at least one public lecture in Oil Springs in January 1866.[80]

The relationship between science and religion preoccupied Winchell in the late 1850s and into the 1860s, when he began to tour giving popular lectures on scientific topics. In *Sketches of Creation* Winchell synthesized science and theology using evidence drawn from North American geological reconnaissance. "Science interpreted is theology," Winchell writes. "Science prosecuted to its conclusions leads to God."[81] He attempts to reconcile the six days of the Genesis creation story by making each day of creation represent an epoch of great length. *Sketches of Creation* was representative of the new literary genres of reflective science that emerged during the nineteenth century. As James Secord suggests, these writings were unified by a common debate about geological progress buttressed by scriptural authority.[82] A narrative of progress was central to geology, and Winchell offered a divinely directed story of progress using a succession of "sketches" whereby the earth was prepared for the occupancy of man.

The evolutionary narrative emerged alongside the classic form of the historical novel. Secord explicates how the writing of another nineteenth-century evolutionary narrative, *Vestiges of the Natural History of Creation*, drew on literary practices from journalism, the new sciences of progress, and fiction to recast evolutionary narrative for a popular readership. Readers were "written in" or "conscripted" into the very fabric of *Vestiges* as intimate acquaintances. Narrative voice and prospective readers are immediately united through the use of the pronoun "we." An emphasis on what Secord described as "familiar knowledge" continues throughout the text. Similarly, in the opening paragraph of *Sketches of Creation* Winchell draws his readers in as imagined partners in the process of composition. He begins:

> What is this which I have opened from the solid rock? It has the appearance of a bivalve shell, like a clam or an oyster. I was passing a delightful summer-day amid the romantic scenery of Trenton Falls, and broke from the rocky wall of the deep-cut gorge these unexpected forms. Who has not stumbled upon similar shapes at the foot of some beetling cliff, or washed from the weathered soil of some cultivated field? Pause a moment, for these are remarkable and unexpected discoveries. Let us interrogate these forms.[83]

The intimacy established in the introductory paragraph is reinforced in the chapters that follow. Winchell continually engages the senses of the readers as the narrative of creation unfolds.

The main organizing metaphor is the ancient one of the "book of creation." As Winchell proceeds through the stages of creation, beginning with "the ordeal by water," through the "ordeal by fire," and the creation of the solar system, he establishes God's role as the creator. Winchell writes that "the whole plan of creation was mapped out to the mind of the Creator in the beginning." This Winchell asserts will become clear to the reader as the story unfolds. He continues, "We shall see, as we proceed with our sketches of the history of creation, that every step in the evolution of continents, and the establishment of a home for the coming man, was a movement in a definite direction, effected by forces chosen from the first, and shaped always with references to exigencies which were to arise in the far-distant future."[84]

In the progressive narrative culminating in the creation of man, Winchell reasons that vegetable life preceded animal life, as vegetable life is capable of enduring more extreme conditions. Winchell cites the general distribution of petroleum down to the primeval gneissic strata as evidence supporting the doctrine of praezoic vegetation, although he also indicates that geologists are divided as to whether animal or vegetable organisms afforded most of the native oil.[85] A chapter of *Sketches of Creation* is devoted to geological theories on the general conditions of oil accumulation and its exploitation. "The smell of the article has turned men crazy," he writes. "It has opened the purse-strings which the cries of the orphaned, the tears of the widowed, and the pleas of religion could never loose." Economic motivations caused men "to scorn the admonition of the instructed and the professional, to trust their own start ignorance in the stake of a fortune." "No doubt," Winchell states, "many men made suitable recognition of the services of the Almighty in facilitating the ends of money-getting."[86] In this chapter, Winchell weaves the tale of "oil fever" as it raged over the Northern United States and Western Canada from 1862 to 1866, which he personally witnessed from a professional standpoint, with the various geological theories about the accumulation of oil and the extent of the supply.

Winchell reviews the various theories on the origins of oil. He refutes the popular theory that the origins of oil lie in the coal beds. Next, he challenges contemporary theories of the origins of petroleum using observations drawn from his fieldwork in Enniskillen

Township. He dismisses the premise put forth by Hunt that Corniferous limestone was "the mother-rock of large supplies of petroleum." On the contrary, Winchell theorizes, "if it were the source of such marvelous quantities as have been drawn from the Canadian strata, its own cavities and interstices should certainly be charged with the liquid." He based his conclusions on the deep "test well" bored at Enniskillen in 1863, which, at a depth of some two hundred feet into the Corniferous limestone, was abandoned. "From the time of entering it the signs of oil were materially diminished instead of increased," Winchell remarks.[87] Instead, Winchell suggests that the Marcellus shale in the lower portion of the Hamilton formation was the primary source of the petroleum found in the Enniskillen and Bothwell regions in the 1860s.

Winchell describes the flowing wells of Enniskillen Township beginning with the Shaw Well struck on 11 January 1862, and he incorporates Hunt's anticlinal theory of oil accumulation into the narrative of evolution. "The rocky retorts in Nature's vast laboratory are warmed," he writes, "their organic contents undergo a slow distillation – the products escape in the form of gas or oil, and slowly filter through pores and crevices toward the surface, till intercepted by some impervious stratum."[88] Adhering to the evolutionary narrative form he writes, "that before October not less than thirty-five wells had commenced to drain a store-house which provident Nature had occupied untold thousands of years in filling for the uses – not for the amusement – of man." In a separate appendix Winchell provides a detailed record of the depth and yield in barrels of the flowing wells of Enniskillen compiled from his own personal examination and research.[89] Using his observations in the field, Winchell calculated that no less than five million barrels of oil floated into Black Creek and ignited. He cautioned "the Canadian squatter of the danger, no less than the utility and wastefulness, of his oleaginous pastimes."[90] In Winchell's creation narrative, the beneficent provisions of the earth's crust are intended for man's use alone.

How widely *Sketches of Creation* was read by the settlers and oil speculators in Enniskillen Township is unknown, although it was almost certainly read by the emergent middle class of the fledgling oil resource town of Petrolia during the 1870s. Alexander Winchell was well known in the region through his geological fieldwork and his public lectures. James Kerr, a local oil producer, lumber merchant, and avid amateur geologist, referred to *Sketches of Creation* in a feature article entitled

"The Oil Belt," published in the Toronto *Mail* on 1 December 1888. The article reveals something about how the book was read locally during the latter nineteenth century. Kerr drew specifically on the chapter of Winchell's book that focused on the output of the flowing wells during the 1860s, and how much of the oil was wasted because of inadequate storage.[91] Strangely, given that Kerr probably read widely in the field because of his interest in geology, the broader narrative of creation articulated by Winchell was not mentioned in the article, nor was the argument about the relationship between religion and natural science.

With the cessation the flowing wells at Oil Springs early in 1863, the question of the extent of the Enniskillen oil fields was never far from the minds of speculators. "Wildcatters" and practical men in the field looked askance at professional geologists who frequently failed in their attempts to locate productive fields. The wildcatter's drill frequently found oil in locations at variance with the anticlinal theory of geologists.

Another character type, known variously as the "dowser," "oil smeller," "witcher," or "doodle-bugger" appeared in the nineteenth-century search for oil in North America.[92] In *Sketches of Creation*, Alexander Winchell indicated that "oil smellers in high ecstasies" were seen in Northwestern Pennsylvania, "men whose experienced olfactories were employed to test the odor of every bog, and stain, and film which preying eyes could bring to light." Winchell described them as having "a wisdom above geology." In many instances the market value of property tripled with "the magic touch of the magician of the hazel wand."[93]

In April 1865 a correspondent for the *Globe* reported that a "witch" had been encountered in Enniskillen Township searching for oil.

> You have read how, in the old German fatherland, ere Hengist and Horsa had crossed the German Ocean, and how in Scotland and in England, in later days, the split or forked willow stick was used as a divining wand. So it is in this present day in Enniskillen, Canada West. There is a man here, on competent authority pronounced to be a "witch" – the masculine wizard being unknown in this locality – who will take a willow branch in his hand, and by it action tell to a certainty where oil will be found if the ground be bored ... The wizard chooses a forked branch, the stem of which he holds tightly in a horizontal position. And thus he marches in stately fashion over the ground, until, by a power to him irresistible, a branch of the fork is drawn down in the direction

of the "ile" vein. When once there the strongest man in Enniskillen might strive in vain to move it. It will not budge.

"There are some sceptics here who laugh at the "superstition," as they are pleased to call it; but there are plenty ready to testify strictly to the truth of the facts set forth," the reporter remarked.[94]

The birthplace of the modern use of the dowsing-rod was in the mining districts of Germany. It was probably introduced into England in the latter part of the sixteenth century by German miners hired to work in the Cornish mines. At the end of the seventeenth century, the ability of a dowser to locate water or minerals using a divining rod was usually attributed to some psychic or spiritual phenomenon.[95] In the cultural construction of science and the development of the oil industry in nineteenth-century Enniskillen Township the work of the "oil smeller" was defined as "magic" rather than as a scientific "fact."

The *Sarnia Observer* described an oil smeller who wandered into Enniskillen Township in December 1865 as an "antique specimen of humanity." "Clad in the most dilapidated garments, face furrowed by exposure and mental conflicts, beard matted and gray, he looked like a wanderer under the ban of some despotic power," the *Observer* reported. "Tied up in part-colored rags were his wands and rods, and various sized phials containing dark fluids. A queer apparatus connected with a bottle of quicksilver, he styled 'the confirmer,' and its power, he asserted, was indisputable 'in the dark of the moon.'" The oil smeller's fee for locating a well was reportedly five dollars, with a guarantee that oil would be discovered within one year.[96]

Another correspondent for the *Observer* described the technique used by an oil smeller who visited the Enniskillen territory:

> The operator proceeds across the fields, or along the bank of a stream, or tracing up a ravine, and occasionally pauses, standing on one foot, fixing his eyes earnestly on the ground. While both feet are on the earth, the theory is that a magic circle is formed, and the same sensation is not felt in the nervous system as when the connection is broken by resting on one foot only. Hence the magnetic sensation is all passed through one limb, and whenever the oil is beneath, the feeling is greater or less in force according to its proximity with the surface, or its extent of volume.[97]

Oil smellers were a curious attraction in the nineteenth-century Enniskillen oil fields. During the 1860s, professional geologists, practical men, and wizards all met in the oil fields of Enniskillen Township where they articulated their ideas about the origins, location, and the extent of the oil resources in the region. "All this," Alexander Winchell wrote in *Sketches of Creation,* "because 'oil' is a synonym for gold."[98]

CHAPTER THREE

The Petrolia Discovery and the Making of an Oil Resource Community

In 1880 *Belden's Illustrated Historical Atlas of the County of Lambton* asserted that "Petrolea is essentially and entirely an oil town – not to be compared, perhaps, with some of those in Pennsylvania oil regions, yet an oil town, strictly and literally, notwithstanding." Belden built on its characterization of Petrolea as an "oil town" as follows:

It had its inception in oil. Its early stages of devolopment [sic] were fostered and supported by oil. Its present prosperity and activity is the direct result of the production of oil. And it is claimed that even its future existence is dependent upon the continued profitable yield of oil. In fact, the alpha and omega of Petrolea is oil, oil, oil. Everything smells of oil; everything tastes of oil; everything is covered and smeared with oil; everything *is* oil. You hear nothing but oil spoken of in the cars, in the hotels, in the public offices, in the stores, in the "Exchange," on the streets, *everywhere*; and we would think from a casual visit that not only the prosperity of Petrolea, but the lives of all its inhabitants and the existence of the whole country, depended on whether "crude" advances or declines one-eighth of a cent "on 'Change" within the next ten days.[1]

A correspondent from the Toronto *Mail and Empire* echoed the depiction of Petrolia presented in Belden's Atlas. Visiting Petrolia in the spring of 1899, the reporter observed: "One does not need to remain in the town long to find that Oil is King, and that the place depends upon it almost completely." The inevitability of oil production surrounds the place, the reporter indicated, and wherever a labourer was able to save up enough money an oil well displaced the flower garden in the front yard.[2] The oil industry provided the expression and precise definition of communal boundaries that shaped class identities in Petrolia from about 1861, when oil speculators turned their attention to the oil deposits located on the flats of Bear Creek, to 1898, when the U.S. conglomerate Standard Oil took over Imperial Oil and transferred their refining operations to Sarnia.

Communities, as the writings of cultural critics have taught us, are practices of collective identification. Boundaries are constructed and maintained through the shared discourses of everyday life. British Marxist literary critic Raymond Williams rejected the stereotypes associated with modernization theories that set up a dichotomy between the city and the country, with the city as a place of noise, worldliness, and ambition, and the country as a place of backwardness, ignorance, and limitation. Small towns, Williams points out, contain many intelligent, strong, and honourable men and women, and it is an injustice to see them as backward and ignorant. A knowable community, both in the country and the city, is a matter of consciousness and continuity, as well as day-to-day experience. Williams articulated an idea of "active community," where a new kind of community only became a reality when economic and political rights were fought for and partially gained.[3]

Recent research in the area of cultural geography suggests that communal boundaries are also constructed in socio-cultural space.[4] Boundaries are constructed and maintained through the discourses of everyday life. Community boundaries also incorporate and enclose difference. "Insiders" shared access to a common discourse built around the oil industry that was unfamiliar to "outsiders," to whom access was denied. Discussion about the conditions in the oil industry included *all* Petrolians, not just oil producers and refiners.

Historians of class formation have routinely, albeit sometimes implicitly, incorporated socio-cultural space into their analyses by identifying men and masculinity with the public sphere of politics and commerce, and women and femininity with the private sphere of home and child-bearing. In small and medium-sized towns throughout nineteenth-century Ontario, as Andrew Holman and David Burley explain, middle-class formation occurred as the emergent "middling orders" of businessmen, professionals, and white-collar workers began to transform their respective communities from pioneer settlements into bustling "go-ahead" locales.[5] The character of the middle class resonates with place, and was shaped by local circumstances and conditions. A man's individuality and his masculine gender identity were closely tied to his independence and his occupation. The formation of a male Anglo-Protestant middle class of prosperous "oil barons," merchants, manufacturers, and professionals was vital to the transformation of Petrolia from a crude, rough-cast, mining village into a respectable middle-class Victorian community. The "town fathers," who were also oil producers, refiners, and merchants, carved out a civic space that was created and maintained as a middle-class masculine province. Middle-class men's power in the economic realm translated into power in the public sphere of local politics and male associational life. The Anglo-Protestant, male, middle-class public sphere in latter nineteenth-century Petrolia, as will be discussed in chapter 4, was constructed on the basis of exclusions, most notably those of class, gender, ethnicity, and religion. Two nineteenth-century masculine gender stereotypes were interwoven in the formation of community and class identity in Petrolia, as evidenced in the character of the risk-taking, hard-drinking, assertive, and vigorous oil speculator usually associated with the rough culture of the working-class tavern, and the hard-working, disciplined, and enterprising businessman.[6] Neither type was exclusive of the other. Individuals sometimes possessed characteristics of both gender ideals. A small group of risk-taking oil producers remained in Petrolia, started families, and rose through the ranks to become part of the prosperous middle class that transformed Petrolia from a crude shantytown into a respectable Victorian oil community.

THE PETROLIA DISCOVERY

Charles Tripp located oil deposits on the flats surrounding Bear Creek as early as 1853. The Tripp brothers acquired various properties on the tenth and eleventh concessions, but the flowing wells at Oil Springs attracted the attention of oil speculators and fortune seekers. A natural oil spring was allegedly discovered sometime in 1860 or 1861, although there is some discrepancy as to its exact location. A correspondent from the Toronto *Globe* wrote in September 1861 that the fields were located on lot 13, concession 11, whereas several contemporaries and local historians reported that the spring was located on lot 13, concession 10, on the farm of John Bligh or Blyth.[7]

With the renewed interest in oil speculation in Enniskillen Township in the spring and summer of 1861, newspaper reporters flocked to the area. They provided some of the earliest descriptions of the emergent oil resource community around Bear Creek. At a distance of about four miles from the village of Wyoming they came upon a store and tavern. "This is Petrolia," a reporter from the Toronto *Leader* wrote: "West from it about half a mile are the wells known as "Kelly's Wells." Here, he noted, about a dozen wells were "yielding tolerably."[8] Another reporter reckoned that there were a hundred wells, at various stages of completion, along the banks of Bear Creek, employing about 300 men; the most successful of them were owned by Mr Kelly.[9] A refinery had just been completed and had commenced operations. The refinery, owned by A.A. Adams & Co. of Boston, covered three acres on the flats of Bear Creek. The refined oil produced by Adams & Co. was allegedly of the best quality, but there was a great deal of waste. No use was made of the benzole that was allowed to escape, and the refuse seeped back into the creek.[10] The *Globe* reported that in comparison to Oil Springs eight miles to the south, "Petrolia is comparatively a pretty place." "There are extensive clearings about it," the *Globe* revealed, "farm houses may be seen here and there upon the ridges of the rolling land, and Bear Creek, upon the banks of which the wells are situated, does not smell of oil, and is as clear a stream as Black Creek is a dirty one." The population of Petrolia was estimated at about five hundred.[11] Oil producers, nevertheless, continued to concentrate their capital investments on the Oil Springs field until the widespread cessation of the flowing wells occurred, beginning in the spring of 1863.

While newspaper accounts referred to the settlement as "Petrolia" in the spring of 1861, the naming of the community has been attributed to Patrick Barclay, the first postmaster. Barclay and three others allegedly named the community "Petrolea" in 1861.[12] Although the spelling "Petrolea" was retained for many years, a clerical error when the village was incorporated as a village in 1866 resulted in a change in spelling to "Petrolia," although both spellings were used.

By the spring of 1865 the settlement at Petrolia consisted of six log cabins. The collapse of the oil industry at Oil Springs in 1863, together with an increase in the price of oil on the market, caused speculators to turn their attention once again to the Bear Creek wells. Loi Reduc, the local correspondent to the *Sarnia Observer*, disclosed that nineteen acres of land had been sold at Petrolia for the sum of $21,000 in gold, and prospecting was to begin forthwith. "The Michiganders have been rolling in by the load," he wrote. "Pontiac, and vicinity, Flint, Adrian, and other conspicuous places of the Peninsular State, have been represented by inquiring capitalists and secretive gents, who live without labor." At first the Petrolia field attracted small-scale speculative capital intent on making a quick return for their investment. Many of these oil speculators were from the U.S. They moved northward from Michigan and the Pennsylvania oil fields in the aftermath of the Civil War. The masculine gender identity associated with the American oil speculator was that of the risk-taking gambler. He was viewed with suspicion, and stories filled the pages of the local press focusing on unscrupulous American oil speculators who swindled innocent small producers seeking their fortunes in Enniskillen Township. For instance, Loi Reduc wrote:

> Here is a man who attends to his own affairs scrupulously, and avoids society. He is adjudged as a behind-the-door drunkard. Another of a nervous temperament is always in a hurry. He has done something wrong and is trying to conceal it. A third is a frequcnter of the hotels, and it is a pity that his employers do not know how shamefully he is neglecting their affairs.

The reporter noted that social scandal permeates the region and "it is merciless as regards sex." He continued, "A lady promenades the single sidewalk at twilight, and her husband ought to know something about it. Or if seen gathering moss in the edge of the bush, the observers thought she was queer, to say the least of it." The respectability of women in public was suspect in this rough mining community. Loi Reduc advised anyone thinking of visiting the region to "[l]eave your sensibilities at home when you come here, whether for a day or a week."[13]

Tales of innocent gentlemen duped by oil speculators were components of the culture that sprang up around the Petrolia oil fields. Oil speculation was satirized in a short three-act drama published in the *Sarnia Observer* on 30 June 1865, entitled, "Petroliana: A Warm Weather Sensation." The names of the main characters were derived from the rhetoric of the oil fields. *Ded Beet, Ben Zeen,* and *Kerro Seen* set out to dupe an innocent gentleman, named *Verdi Green,* into purchasing mining stock on a piece of property on lot 23, concession two, which is unlikely to produce any oil. The oil speculators concoct a story about a man who dug for water on the property six years earlier, and at a depth of sixty feet was forced to stop drilling to fish for a crow bar. As he attempted to pull out the crow bar, oil rushed to the surface. While the drilling crew was devising a scheme to save the oil, the ground trembled and the well caved in. *Hazel* enters the room and informs the speculators that a man wanted to purchase "twenty-three." The men trick *Verdi Green,* who has no knowledge of the local oil industry, into purchasing the property. The drama concludes with an air about the dubious practices of oil speculators sung to the tune of "Camptown Races."

Ben.– Now we've made a fair divide;
Drill down, drill down,
Of twenty-three steer very wide,
Drill down, drill down there;
Suppose they drill all night,
Suppose they drill all day,
The more they drill they'll soon find
The oil has gone away.

Ker.– "Let up" on gold and oil produce,
Drill down, drill down;
There's greenbacks plenty lying "loose,"
Drill down, drill down there;
You want but little cash,
It's "hefty" on the pay
Put up your money in Ded Beet's hands,
And let him have his say.

Haz.– Of mining stocks my friends beware,
Drill down, drill down,
Small dividends they now declare,
Drill down, drill down, da;
Buy stock in a flowing well,
Or one that's pumping gay,
It will pay you well if it don't play out,
So take hold while you many.

D.B.– *(aside)* They're jolly now with rolls of cash,
Drill down, drill down;
All which is naught but bogus trash,
Drill own, drill down there;
My purse is bursting full,

I'll need no longer stay,
  I'll take myself to furrin parts,
With a hip, hip, hurrah!
    (They dance – curtain descends slowly.)[14]

Twenty-five acres on lot 14, concession nine, changed hands in late November 1865 for $250 an acre, with the seller realizing a profit in the transaction of approximately $5,000. Fortunes were made and lost in oil and land speculation. Wells were almost always represented as yielding far more than they actually produced. Machinery and skilled workers were in high demand. The majority of owners of oil properties both at Petrolia and at Oil Springs sank their oil wells with their own hands. In the early 1860s a group of skilled oil drillers who accepted contracts to bore to any depth required arrived. The average rate was $2 per foot for the first one hundred feet; $3 per foot for the second hundred feet; and $4 per foot for the third hundred feet, with each one hundred feet drilled requiring an additional man to work the drill. A steam engine was usually used after a depth of three hundred feet was attained. With the demise of the Oil Springs field, the Petrolia fields began to attract more large investors from the U.S. The machinery used in the Petrolia fields was judged superior to that used in the Black Creek fields, and allowed drillers to bore deeper into the rock, although breakdowns occurred frequently.[15] The costs of drilling in the Petrolia fields far exceeded that of the Oil Springs wells.

Accommodation in the fledgling oil resource community was at a premium. Loi Reduc complained that the hotels were not ready for guests on Christmas as anticipated, and he suggested that "[t]here will be competition enough to ensure good accommodations for transient or permanent patrons."[16] Two "experienced" parties from Titusville, Pennsylvania, were rumoured to be opening an "oyster bed;" "so that when it is short commons at the hotels, the appetite may be appeased by disbursing an extra quarter outside."[17]

In 1865 John H. Fairbank relocated from Oil Springs to Petrolia. During the next seven years he established himself as an oil merchant, real estate speculator, banker, retail merchant, and politician. Upon his arrival in Petrolia, Fairbank went into business as a general retail merchant. Fairbank's general store began as a grocery and liquor store, and local farmers made their exchanges there, but he soon expanded into hardware and oil-well fittings. In 1869 the business was known as Fairbank & Bennett. Fairbank's partner, Alexander B. Bennett, probably managed the store. In 1874 the business was renamed VanTuyl & Fairbank. Fairbank's partner, Major Benjamin Stoddard VanTuyl, was a veteran of the U.S. Civil War, and the son of a wealthy New York State merchant, lumberman, and landowner. VanTuyl arrived in Petrolia in 1866 and went into business for himself drilling artesian wells and operating oil wells.[18] Above the store was a room known as "Fairbank Hall," which served as a meeting place for oil producers. The market price for crude and refined oil was negotiated between oil producers and refiners at Fairbank Hall before the construction of the Oil Exchange Hall in 1871.

In partnership with Leonard B. Vaughn, Fairbank established the banking firm of Vaughn and Fairbank on 12 August 1869. Vaughn began his banking career in Oil Springs, where he was also an oil producer. Vaughn and Fairbank's bank, known locally as "the little red bank" after the colour of the building, provided an extensive line of credit to small producers during the economic downturn of the early 1870s. One pioneer Petrolia oil producer, Charles Jenkins, remarked that the style of the two bankers suited the place. "They required no mercantile references as to anyone's standing," Jenkins wrote. "The quality we call human nature was largely developed in both partners, and decisions in giving credit were generally determined by their knowledge of the borrower's character and record."[19]

On 23 November 1866 Captain Benjamin King, manager of the Great Eastern Oil Company of St Catharines, brought in a large flowing well located on lot 11, concession 11, in a heavily wooded area nearly covered in water. The well, located by the celebrated oil-smeller, Mr Kelsea of Buffalo, was situated approximately one and a quarter miles north-west from the wells on the flats of Bear Creek. The vein was struck at a depth of 383 feet with an ordinary 4 ½ inch drill, "throwing out a volume of gas and water with a roar that could be heard a distance of 2 miles." "The workmen were nearly suffocated with the gas, and had to be carried out of the derrick," the *Globe* reported. "In about two minutes, the gas and water was followed by a volume of pure oil which was forced completely over the top of the derrick, at a height of 48 feet."[20] At first the drillers were unable to contain the oil, and a pool of oil about 100 yards in length, approximately 12 feet in width, and one to two feet in depth formed. A 250-barrel tank was quickly constructed, into which pumped an estimated 800 barrels of oil a day.[21] King's success astounded oil speculators. Because the oil strikes at Oil Springs had been made close to the course of Black Creek, drillers who ventured to Bear Creek believed that oil could be found only along watercourses, and hence they confined their operations to the creek bed. Drilling operations were quickly transferred from the Bear Creek flats to the new field.[22]

The strategies of resource extraction influenced settlement and community development. The first oil discoveries were to the east, but by 1867 the oil-producing district had spread westward across Bear Creek and covered much of what is now the business district of Petrolia. Like any small nineteenth-century resource town, the first buildings were of wood frame, constructed in the cheapest and quickest manner for instant occupation. They were never intended for long-term use or permanent settlement.

The growth of the community was phenomenal. In 1861 the population was 500; five years later it was nearly 2,300, although the population fluctuated in the intervening years owing to fluctuations in the oil market. On 13 December 1866 Petrolia was incorporated as a village, and later as a town in 1873. The success of oil producers, merchants, and professionals in business translated into moral authority in the public realm and power in local government. The councillors of the new village were young men. The Reeve, William H. McGarvey, was only twenty-four years old. The oldest councillor, postmaster Patrick Barclay, was thirty-nine years of age. The enterprising young businessmen and oil producers who controlled the civic space promoted the village's growth and prosperity as a way of furthering their own business interests. Yet their success in business endowed them with the skills necessary to provide leadership as the oil resource community took shape. After assisting his father in the family grocery business in Wyoming, William McGarvey relocated to Petrolia in 1861. After serving only a few weeks he resigned his position as Reeve to open a general store, which he named "The Mammoth Store." McGarvey also secured an interest in several oil properties, including the famous "Deluge" well struck in 1873, and a partnership in an oil refinery. He was elected Mayor in 1876 and Warden of the County of Lambton in 1879. In the 1880s McGarvey transported the Canadian method of "pole" drilling to "foreign fields."[23] George Moncrieff, a young lawyer, who was also heavily invested in oil production, replaced McGarvey as Reeve. Although these enterprising businessmen were committed to promoting their town's prosperity, they also looked to further their investments in the oil industry. As was the case in other small towns eager to promote manufacturing, tax concessions were routinely

Figure 3.1 Petrolia in 1866. Lambton Room, Lambton County Library, Wyoming, Ontario

made to oil refiners over the next two decades to promote growth of the industry during a period when the country was engulfed in a deep depression.

The attraction of the Petrolia oil fields was their perceived permanence. By the spring of 1866, one well, the "Eureka," had been in operation for over fours years. The village of Petrolia developed as a small service centre for the oil industry. A main street with nine hotels, an oil exchange, and a telegraph line emerged over a four-month period, and retail businesses lined the west end of the village near the wells on the Bear Creek flats. As local historian Victor Lauriston noted, the names of these hotels – the Great Western, the United States Hotel, the American Hotel, the Michigan Exchange, the International, the Saginaw House, and the New York House – point to the U.S. origins of many of the first oil developers in Petrolia.[24] McEvoy's *Gazetteer* for 1866 noted: "There are several large and well-stocked stores, and others are being rapidly built." Four churches – Roman Catholic, Canada Presbyterian, Wesleyan, and Episcopal Methodist – were built near the flats.[25] McEvoy's *Gazetteer* further stated that "The rapid rise of Petrolia, and the unparalleled success of the wells which have been sunk since oil was first discovered here, have prevented, in a great measure, the several improvements requisite in a village of its size ... the early gold seeker to California or to Pike's Peak will be reminded here of his experience in those Pacific-bound localities at the time of their first discovery. Situated in the midst of a woody country, with hill, dale and swamp in abundance, it will be some time before the village of Petrolia will present the appearance of an old established settlement."[26]

The great difficulty faced by the oil producers was how to store the oil and convey it to the Great Western Railway line at Wyoming. In 1862 a survey was made for a branch line from Wyoming to Oil Springs and in 1863 a charter was secured for the work. An alternative route from Newbury on the Toronto-Windsor line to Enniskillen was also considered. The directors of the railway were reluctant to invest in a railway line to Petrolia, deciding instead to wait and see whether the oil field proved productive and enduring. No action was taken until 1866, when George Lowe Reid, Chief Engineer of the Great Western Railway, purchased the right-of-way and encouraged other company men to invest in a spur line from Wyoming to Petrolia. Local historian Edward Phelps suggests that J.H. Fairbank was probably largely responsible for the organization and financing of the line. In December 1866 a lavish dinner was held in the United States Hotel to celebrate the opening of the Petrolia spur line.[27]

Fire was always a concern for the residents of Petrolia. On 25 July 1867 fire broke out near the King oil well. The fire spread to the adjacent field, owned by John D. Noble, on the north-east corner of the west quarter of lot 12, concession 11.[28] The fire consumed Noble's storage tanks and an estimated 7,500 barrels of oil. The fire spread over twelve acres, with flames shooting as high as 100 feet. Noble recalled that it took nearly two weeks to extinguish the fire. A second fire broke out on August 2 when a maintenance worker inspecting a storage tank held his lantern too close to the stream of gas rising off the crude. Before the fires it had been the practice to store surplus crude in surface tanks, mostly made of wood, but some were made of metal. "I thought over the matter how a recurrence of such a fire could be avoided, and I devised the tanking system," Noble later recalled in his testimony before the Royal Commission on the Mineral Resources of Ontario. "The clay here is of such a nature that it makes a perfect tank in itself." "On top of the clay there is an alluvial deposit of about 15 or 18 feet," Noble explained. "Excavations are made about 60 feet deep and 30 feet in diameter, with a capacity of 8,000 barrels of 35 imperial gallons. When we get down 18 or 20 feet, to the clay, a wooden sheathing is put round the tank, clay is pounded down firmly, and the wall is brought up round the tank, clay is pounded down firmly, and the wall if brought up about a foot above the surface. At the top it is sloped off so as to prevent water getting into the tank, and it is perfectly water-tight."[29]

The population of Petrolia during the 1860s was transient and unsettled. Like the nineteenth-century mineral resource towns of western Canada studied by Cole Harris, the rhythms of work and leisure were bound to the technologies of resource procurement, and economies were tied to fluctuations in the oil market.[30] In January 1867 the *Sarnia Observer* reported that the oil business was exceedingly dull and that nearly all of the wells were shut down. The decline in the price of oil reverberated throughout the local economy. According to the *Observer*, few wells were going down, and any new wells were paid for with interests in property. Almost no currency was circulating in Petrolia and business was at a standstill. As a result, merchants who had given credit found it next to impossible to collect. Also, the floating population of transient workers had been reduced and hotel keepers suffered from the decline in visitors and regular boarders, many of whom simply disappeared without settling their bills.[31] The problem of a market for Canadian oil also plagued oil producers during the late 1860s. An impediment to the introduction of Canadian oil into the European market was its strong odour. American oil was more easily marketed abroad as it was less odoriferous.[32]

OCCUPATION, SPACE, AND CLASS FORMATION IN PETROLIA, 1871–91

When the census was enumerated in the spring of 1871, the population of Petrolia was 2,651. As Table 3.1 indicates the ratio of men to women was 1.2 to 1. Surprisingly, given the numerous newspaper accounts referring to the U.S. origins of many of the oil speculators, only 7.5 per cent of the residents reported the U.S. as their place of birth. The vast majority of the residents, 70.9 per cent, were Canadian-born. Those born in England comprised 9.1 per cent of the population, and those born in Ireland and Scotland, 5.5 per cent and 6.4 per cent respectively (see Table 3.2). As Table 3.3 suggests the majority of residents were Protestant: 36.3 per cent Methodist; 26.3 per cent Presbyterian; and 25.8 per cent Church of England. Only 5.9 per cent of the population reported Roman Catholic religious affiliation. By 1881

Table 3.1 Population and rate of growth, Petrolia, 1871–1911

Year	*Population*	*Growth Rate*	*Ratio of Men per 100 Women*
1871	2,651		121
1881	3,465	31.0%	105.6
1891	4,357	26.0%	102
1901	4,135	-5.1%	92
1911	3,518	-15.0%	94

Source: Canada, *Census of 1871*, vol. 1, table 1; Canada, *Census of 1881*, vol. 1, table 1; Canada, *Census of 1891*, vol. 1, table 1; Canada, *Census of 1901*, vol. 1, table 1; Canada, *Census of 1911*, vol. 1, table 1

Table 3.2 Birth places, Petrolia, 1871–81

	1871		*1881*	
	no.	%	*no.*	%
Canada	1879	70.9	2646	76.4
England	240	9.1	837	8.4
Ireland	142	5.5	170	4.9
Scotland	169	6.4	145	4.2
U.S.	198	7.5		

Source: Canada, *Census of 1871*, vol. 1, table 2; Canada, *Census of 1881*, vol. 1, table 2; Canada, *Census of 1901*, vol. 1, table 2

Table 3.3 Religious affiliation, Petrolia, 1871–1901

	1871		*1881*		*1891*		*1901*	
	no.	%	*no.*	%	*no.*	%	*no.*	%
Baptist	76	2.9	358	10.3	348	7.9	594	14.4
Roman Catholic	759	5.9	204	5.9	229	5.2	304	7.4
Church of England	685	25.8	853	24.6	653	14.9	631	15.3
Methodist	962	36.3	1234	35.6	1212	27.8	1456	35.2
Presbyterian	698	26.3	736	21.2	821	18.8	1030	24.9
Salvation Army					51	1.1	65	1.6

Source: Canada, *Census of 1871*, vol. 1, table 4; Canada, *Census of 1881*, vol. 1, table 4

the population of Petrolia had increased to 3,465. In 1881 the majority of the residents, 76.4 per cent, reported Canada as their place of birth, and most were Protestant.

The census figures, however, mask the decline in population with the depression of the mid-1870s. The population of Petrolia leaped from 2,651 in 1871 to 4,239 in 1872, only to drop sharply to 3,667 in May 1873. The population declined to a low of about 2,346 in 1875 before rebounding.[33] Between 1871 and 1891 the population grew by more than 25 per cent each decade. In 1891 the population of Petrolia was 4,357. In 1901 the ratio of men to women reversed, and there were slightly more women then men (see Table 3.1). Prompted by the relocation of Imperial Oil to Sarnia in 1898, the population of Petrolia began to decline, and in 1901 the population was 4,135.

Contrary to the early representations of Petrolia as a community of rough, risk-taking, single men, by 1871 the village already comprised mostly single-family households. Of the 481 households enumerated in 1871, men headed 471 households and women headed 10 households. In 1881 a total of 668 households were enumerated: 632 were male-headed and 36 were woman-headed. In 1891 the total number of households had increased to 878, with 828 headed by men and 58 headed by women. The majority of household heads, both male and female, who reported an occupation were identified as working class.

From the 1871 census of manufacturers and from various business directories, few industrial sectors could be identified that were not in some way connected with oil production or serviced the oil industry. In a small shop, where he employed one man, Samuel Carr combined shoe and boot making with oil valve repair work and the manufacture of seed bags for drill packing. Blacksmiths and machinists manufactured oil-well casings and drill bits and carried out repair work on drilling equipment. Tronson Draper, an oil producer, set up a machine shop in Petrolia in 1866 to manufacture steam engines and drilling equipment. In 1871 Draper employed six men. His business grew and beginning in the early 1880s equipment manufactured in Draper's machine shop accompanied Enniskillen drillers to "foreign fields."[34] James and George McGill employed eight men in their cooper shop producing and repairing oil tanks and oil barrels. Another one of the village's large employers, Thomas Penton, employed twelve men and two boys in his shop manufacturing oil-well casings and battens. In 1871 James and John Kerr employed twelve men in their carpentry shop, where they processed lumber for dwellings and constructed rigs for oil wells. Table 3.4 reveals that the majority of the manufacturing establishments enumerated in Petrolia in 1871 were small shops, employing fewer than five workers.

The oil refineries were among the village's largest employers. In 1871 the Canadian Land and Mineral Company employed twenty men, and produced 588,240 gallons of oil valued at $90,000. Established in 1869 under the management of John McMillan, the refinery

Table 3.4 Industrial establishments in Petrolia, 1871: Number of shops, number of workers, and average number of workers per shop by productive sector

Sector	*A Shops*	*B Workers*	*B/A Workers per Shop*	*Percentage of Workers in Shops of Five or Fewer*
Metal	14	54	3.9	50.0
Wood	5	29	5.8	10.3
Transportation Equipment	2	2	1.0	100.0
Clothing	8	24	3.0	50.0
Leather	6	14	2.3	100.0
Food	3	8	2.7	100.0
Photograph Gallery	1	0	0.0	0.0
Printing Office	1	3	3.0	100.0
Petroleum Refineries	6	56	9.5	14.0
Total	46	191	4.2	39.8

Source: Canada manuscript Industrial Census of 1871, mfm. copy

was capable of turning out 1,200 barrels of refined oil per week. The company also set up a tin factory to produce five and ten gallon cans with the intention of marketing the refined oil in the European market.[35] The Britannia refinery employed ten men and produced 4,800 barrels of refined oil and 2,000 barrels of tar valued at $39,000. In 1871 the Ontario Carbon Oil Company, also established in 1869, employed nine men and produced 58,968 barrels of distillate. The company erected a mammoth still capable of distilling 2,000 barrels of crude oil directly from the wells by means of powerful steam pumps. The still was capable of producing nearly 1,000 barrels of fine white distillate per week, which was then treated in Hamilton for overseas markets.[36]

In Petrolia shared knowledge, developed around the oil industry, guided and informed the everyday behaviour of the residents. In May 1870 the *Sarnia Observer* reported that the oil trade was in a healthy condition, "having resolved itself from a wild speculative business into a regular honest legitimate enterprise, both as regards the production, refining, and other parts connected with the producing or handling of the article." Capitalists in "good standing" were investing their money in developing the territory and building refineries.[37] The following year the members of the Oil Exchange erected a new hall in the west-end of the village opposite the railway station as a public market and meeting place to replace their old gathering place above Fairbank's store. Fluctuations in the price of oil, as determined by the

Figure 3.2 Rows of three-pole derricks, West End Petrolia. Lambton Room, Lambton County Library, Wyoming, Ontario

members of the Oil Exchange, had an impact throughout the community, and all of the residents paid close attention to the activities on the "Exchange." Oil producers discussed prices and made business deals. The Oil Exchange Hall became a focal point for the community, and a public space for the expression of middle-class masculine identity. The Oil Exchange Hall was also a space for community social events and public meetings before the construction of a town hall in 1889. In June 1871, not long after the construction of the Oil Exchange Hall was completed, a ball was held which was attended by about seventy couples of the *"elite"* of Petrolia.[38]

Space was fundamental to the mapping of class structure onto the social and political life of nineteenth-century Petrolia. By the early 1870s, place images took on a prescriptive nature and spatial metaphors were used to express class divisions. In a description of the "Progress of Petrolia," the *Observer* revealed that "close to the Village, a little west of the G.W.R. Station stands the suburb of Pithole, with its innumerable "rigs," and the shanties, the abodes of the labourers employed at the various wells in the vicinity." The name "Pithole" was apparently derived from the deep pits sunk into the heavy clay for oil storage in the late 1860s. The *Observer* pointed out the contrast between Pithole and the "aristocratic quarter" of Petrolia, designated "Quality Hill," where the residences of some of the "magnates" of the village and "the oil princes of the region," were located.[39] Henry Warren Lancey purchased 190 acres of land, five steam engines, rigs, tubing, and seven dwellings from the Crescent Petroleum Association in July 1871 for $14,050. Lancey, the son of a New England merchant, was a wholesale hardware merchant in Portland before relocating to Petrolia in 1865, where he became one of the large oil producers. He also invested extensively in land, and in 1881 he opened a general mercantile business which he operated for seven years before failing health forced him to give up the business.[40] The layout of the residential area, called Crescent Park after the Petroleum Company, but generally known as "Quality Hill," was intended to replicate a New England village street pattern. The surrounding homeowners were responsible for the central park, which was used by the residents of Quality Hill for croquet and cricket. The streets of Crescent Park, Emma and Ella, were named after Lancey's daughters, and another, Emmeline, had the middle name of Lancey's wife.[41] Among the nineteenth-century residents of Crescent Park were: The Rev. J. McRobie, an Anglican clergyman; oil operators Charles Jenkins, W.K. Gibson, James Beresford, and David McKenzie; retail bookseller and stationer William Lowery; lawyers S.F. Griffiths and George Moncrieff; and accountants J.M. Fowler and G.S. Pitikin.

During the 1870s and 1880s a community of "insiders" was formed with access to a common discourse built around the oil industry. In the summer of 1881 a reporter from the Toronto *Globe* visited Petrolia. Comparisons were drawn between the oil resource town of Petrolia and the gold mining communities of California. "Here all the stages of the history of a California gold field have been present," the *Globe* reported, "the rush of population, the wild excitement and mad speculation, the sudden colossal fortunes which have turned the poor man into a millionaire, immense riches missed by the mistake of the moment, and losses which have turned once wealthy men into poor and obscure wanderers over the face of the earth." The *Globe* commented, "Until lately the residents felt inclined to look upon Petrolea much as the Chinese regard California – a good place to make money in, but to be got out of as soon as a sufficiency had been made." A transformation took place in the community, and a middle class of "leading men" emerged. The *Globe* continued, "The later history of the oil fields has been like that of the older gold fields, where mining has parted to a large degree with the element of speculation and the prospects of rapid fortunes, and become a regular branch of industry, with the ordinary risks and ordinary profits of a flourishing manufacturing business."[42] According to the *Globe,* "leading men" with "confidence in the future of the town" were tearing down the small frame buildings on the main street and putting up permanent brick structures in their place.

The *Globe* reporter further described the masculine character traits of the "leading men" of the town. They were "active," "businesslike," and "intelligent"; the antithesis of the stereotype of the "unscrupulous" oil speculators of the 1860s. According to the *Globe,* they possessed "more the air of citizens of a large and busy city than a quiet country town ... Here are gathered men who have seen much of life and understand its ways ... The leading cities of Canada, United States, and England are all represented by good business men, and as a consequence Petrolea can present as fine a body of representative business men as any of the younger cities of the Dominion ... The very class of people are here who will yet make the town all that taste can make it."[43]

The re-imagining of Petrolia as a respectable Anglo-Protestant Victorian middle-class town, in contrast to its origins as a rough frontier mining community, was depicted in the panoramic representations of the town composed by the town's middle-class businessmen and civic leaders. In a special feature entitled "The Oil Belt," published in the Toronto *Mail* on 1 December 1888, James Kerr reiterated that Petrolia was "essentially an oil town," and compared its development to a mining community. Kerr settled in Oil Springs in March 1862, and in 1865 he relocated to Petrolia where along with his brother John he owned a lumber mill and operated a construction company specializing in poles for oil derricks.[44] "Like the majority of mining towns," Kerr wrote, "Its buildings were of frame, and of the most unpretentious description – erected in the cheapest and quickest manner, and bearing evidence of having been put up for instant occupation." The pioneers of Petrolia went there with the intention of returning to their former homes or removing to the city once they had made their fortunes. "But having developed gigantic industries, and formed enduring friendships among their fellows, they at last came to recognize it as a permanent abiding place," Kerr remarked. He continued, "Comfortable and not unfrequently [sic] luxurious residences of brick and

Figure 3.3 Post office and main street Petrolia, 1905. Lambton Room, Lambton County Library, Wyoming, Ontario

wood have been erected, and in several instances handsome brick blocks and commodious industrial structures have been built." In the summer of 1879 banker and merchant Leonard B. Vaughn erected a brick block building that fronted onto the main street. The first floor housed retail establishments and the second floor contained the offices of lawyer George Moncreiff and Dr Lougheed. In 1881 oil producer and banker Lancey decided to re-enter the mercantile business and erected the first large commercial brick structure in Petrolia, valued at approximately $16,000. The structure contrasted sharply with the plain frame shops in which much of the mercantile trade of the town was conducted. Lancey had large stone ornaments cut into the front of the building, and in defiance of the "croakers" who were skeptical about the enterprise he named the building "Lancey's Folly."

Newspaper features focusing on the early development of Petrolia, and the "insider" account by James Kerr, reveal that during the twenty-five years from the early 1870s to the late 1890s, a middle class consolidated in Petrolia. They transformed Petrolia from a semi-permanent rough-hewn shantytown into their idea of a respectable Victorian middle-class town. A man's individuality and his masculine identity were closely tied to his independence and his occupation, while the majority of women remained in the home within a familial frame. Nineteenth-century working-class masculine identities, as the writings of labour historians have taught us, were shaped through wage work where workers were forced to compete with one another for jobs and wages.[45] The "others" to the independent middle-class man, specifically, women, servants, and the working class, were dependent and subjected.

The identification of masculinity with occupation can be discerned in official statistics, such as the decennial census, and in trade directories. The ordering and displaying of status in society has been a recurrent theme in the history of directory publication. As Mary Bond writes, "The directory was and is intended to facilitate communication between buyer and seller, to be a tool for advertising and marketing within a particular community and also to publicize that community and its advantages to a broader audience."[46] Thus the business directory was not a full listing of the population, nor did it include all male heads of household. Males over eighteen years of age, with independent, usually entrepreneurial, involvement in the economy and status in the community, were typically selected. The decennial census further enshrined masculine gender identity with occupation.

The designation of class was made using occupational categories determined by cross-referencing nineteenth-century business directories, census manuscripts, voters' lists, assessment rolls, and local biographies. In most instances the occupational titles derived from the business directories and the censuses gave a clear indication of the nature of the enterprise most men, and some women, were involved in, and provided insight into their relationship to the means of production. As Table 3.5 indicates, a broad definition of "middle class" was used, encompassing the high-status elite and a vast range of those engaged in trade, commerce, and retailing. The middle class was divided into two groups: an upper middle class of professionals, gentlemen of independent means, merchants who controlled large amounts of circulating and finance capital, and manufacturers; and a lower middle class of shopkeepers and small retailers, non-manual white-collar workers, small independent producers with craft titles, farmers with small land-holdings, and individuals providing services and accommodation. Geoffrey Crossick argues that the nineteenth-century lower middle class occupied a position of marginality from the estab-

Table 3.5 Occupational classifications

UPPER MIDDLE CLASS	
1	Professionals: Self-employed, occupational authority, and independence – lawyers, doctors, clergymen
2	Merchants: Control large amounts of circulating and finance capital
3	Manufacturers: Substantial amounts of fixed and circulating capital and the employment of wage workers; includes some large oil producers and refiners.
4	Gentlemen: Independent means
5	Farmers: large land holders, completed farm
LOWER MIDDLE CLASS	
1	Shopkeepers: shopkeepers, retail merchants, dealers, oil dealers
2	White Collar Workers: Non-manual workers, salaried employees, commercial clerks, government employees, professional clerks, teachers, commercial travellers, managers, oil agents, bookkeepers, police constables, editors
3	Master Artisans & Small Manufacturers: Craft titles and small amounts of circulating capital – oil distillers and refiners
4	Farmers: Small land holders, yeomen
5	Services & Accommodation: Hotelkeeper, livery stable owners, billiard saloon owners, Boarding-house keeper
WORKING CLASS	
1	Skilled Workers: Craft titles, workers for a wage, journeymen
2	Unskilled & Labourers
3	Widows

lished upper middle class. "Its members lived and operated within communities," Crossick writes, "and thus in the context of a complex stratification of both the middle class and the working class. They were in contact with a range of social groups, all of which intermeshed within a local society."[47] In the case of nineteenth-century Petrolia, it is difficult to distinguish between oil producers and drillers. Some drillers were contracted or were employed by producers while others worked independently and owned their own derricks, boilers, and engines. Cross-referencing with business directories often illuminated how these men perceived their occupational status in the community, but class identification could not in all instances be established with absolute certainty, nor were the boundaries between the classes always clear-cut. Those with "craft" designations often juggled ownership of a small independent establishment with their own labour. The working class too was divided into two categories: skilled workers, who worked for a wage and who possessed craft titles; and unskilled workers and labourers. In late nineteenth-century Petrolia very few industrial sectors were independent from the oil industry.

Table 3.6 Occupations of heads of household in Petrolia, 1871–1891

	1871 no.	%	*1881* no.	%	*1891* no.	%
UPPER MIDDLE CLASS Professionals	8	1.7	14	2.1	23	2.6
Merchants	8	1.7	18	2.6	1	0.1
Manufacturers	2	0.1	5	0.6	4	0.5
Gentlemen	1	0.2	2	0.3	0	0.0
Farmers	0	0.0	2	0.3	0	0.0
Total	19	4.0	36	5.9	28	3.2
LOWER MIDDLE CLASS						
Shopkeepers	33	6.7	22	3.3	38	4.3
White- Collar Workers	22	4.6	46	6.9	45	5.1
Artisans & Small Manufacturers	102	21.2	101	15.1	127	14.5
Farmers	4	0.8	30	4.2	15	1.7
Services & Accommodation	24	5.0	16	2.4	21	2.4
Total	185	38.3	215	31.9	246	28.0
WORKING CLASS						
Skilled Workers	161	33.5	192	28.6	406	46.2
Unskilled & Labourers	109	22.7	131	19.5	150	17.1
Total	270	56.2	323	48.1	556	63.3
Widows & Spinsters	4	0.8	88	13.2	30	3.4
Unknown	3	0.7	6	0.9	18	2.1
Total	481	100.0	668	100.0	878	100.0

Source: Manuscript, Census of 1871, 1881, 1891, mfm. copy

Table 3.6 reveals that at all three census periods between 1871 and 1891, the upper middle class elite of professionals – gentlemen, large land-holding farmers, and merchants and manufacturers who controlled substantial amounts of fixed and circulating capital – represented a small minority of household heads. The members of this group, which included J.H. Fairbank, James and John Kerr, Charles Jenkins, Leonard Vaughn, John D. Noble, Benjamin VanTuyl, Henry W. Lancey, William McGarvey, and Jacob Englehart, tended to be men of modest means who made spectacular gains. Like their counterparts in other Canadian towns, Petrolia's upper middle class men had various business interests; they were at once oil producers and refiners, partners and independent businessmen, merchants, and land speculators.[48] A case in point is the career of John H. Fairbank. Fairbank was born near Rouse's Point, New York, on 21 July 1831. He immigrated to Canada in 1853, and two years later he married Edna, the daughter of Niagara Falls farmer and United Empire Loyalist, Hermanus Crysler. Fairbank worked on various surveying jobs, and for some time was employed in the office of R.G. Benedict and Ira Spaulding, engineers for the Great Western Railway. In 1861 Fairbank went to Enniskillen Township to survey some bush land that Mrs Julia Macklem had acquired. He hired some men to assist him and surveyed the hundred-acre tract into 198 small plots suitable for sale to oil producers. Attracted by the prospects of oil, Fairbank abandoned surveying and sank his modest resources into oil production and land speculation. He soon owned shares of a producing well and a small refinery. The glut of oil in 1863–64 pushed Fairbank to the verge of bankruptcy, which was forestalled only by having his mother sue him. His fortunes reversed with an upturn in the market, and through steady reinvestment Fairbank eventually became the largest producer in the Ontario oil industry. After relocating to Petrolia in 1865 he diversified his interests into a hardware store specializing in oil fittings and banking.[49]

The Petrolia oil industry depended heavily on small producers with their own sources of capital. The lower middle class largely comprised oil operators and refiners with one or two lots. Numerous small producers, as Hugh Grant argues, made for a highly competitive industry and highly volatile class identification.[50] Fluctuations in the oil market affected the stability of class relationships. Although some oil producers became wealthy men, most of Petrolia's small producers were nowhere near as successful as Fairbank or Jacob Englehart. The boundary between the working class and the lower middle class was always fluid. While there were opportunities for upward mobility, many small oil producers moved back and forth between the ranks of the lower middle class, where they were independent producers, and the working class, where they worked for others as skilled drillers. W.A. Simpson arrived in Petrolia in 1875, and over the next thirteen years he mastered the mechanical art of oil drilling. In 1882 he worked as a driller in the oil fields of Germany. The following year he was employed by the North American Construction Company to sink water wells along the line of the Canadian Pacific Railway. He returned to Petrolia in the fall of 1884 where he was employed by various oil producers to drill in the vicinity of Petrolia and Oil Springs. At various points Simpson was both a small producer and a driller on his own account.

He acquired his own drilling rig and derrick and employed others to work the rig.[51]

Engineers and drillers filled the ranks of skilled workers in Petrolia between 1871 and 1891. Early oil operators would crib down to the bedrock, which lay at a depth of between forty and one hundred feet, and then drill down another three or four hundred feet. In the Petrolia oil field, the spring-poll method was superseded by steam-driven drilling rigs using a poll-tool method of drilling through the rock, thereby reducing the time and expense necessary to dig a well. The jerker system was introduced in the mid-1860s. In this system dozens of wells were pumped using wooden rods connected to a single power source. Oil-well drilling emerged as a distinct and highly competitive business requiring considerable skill on the part of the drillers. As old wells ran dry, they were abandoned, and new wells were constantly sunk.[52] Robert E. Menzie, Manager of the Producers' Oil Refining Company, stated before the Royal Commission on the Relations of Labor and Capital in January 1888 that skilled drillers were sometimes hard to obtain. He stated: "We export about as much labor as any other place in the community. It is in the shape of skilled labor for drilling in foreign countries. Men from here have gone to Germany, Austria, Australia, India and all parts of the world." In the late 1880s, drillers were paid between \$2 and \$3 per day, but they generally worked by the job over a season lasting seven months.[53]

In the late 1880s engineers were paid \$1.50 per day over the entire year. Menzie testified that, "A great number of our engineers are local men, men who have come in here and have run an engine for a short time and get to know the steam gauge, and after they know that and the water gauge they are full-fledged engineers." When questioned about the qualifications of stationary engineers, Menzie replied: "I think a man instrusted [sic] with a boiler, especially when there are persons in the vicinity, should know something about the nature of a boiler and something of his change." He criticized the practice of allowing unskilled men to operate large engines and boilers. Explosions occurred frequently in the refineries and at the oil-well sites. The Royal Commission recommended that a strict system of examination be implemented, and that certificates be granted to properly qualified persons.[54]

Nineteenth-century census gathering practices helped to consolidate the middle-class ideal of separate spheres, whereby women were contained within the realm of kin and family. Census enumerators often ignored women who maintained an independent economic existence. In instances where the home and business was contained within the same establishment the labour of women was also excluded from the census. Dependence was at the core of the Anglo-Protestant middle-class view of womanhood, and the ideal woman was the godly wife and mother. There were few self-employed businesswomen in Petrolia during the nineteenth century. Those women who were self-employed operated within the sphere of women's activities as dress-makers, milliners, and boarding-house keepers. The majority of married businesswomen did not have a husband in the house. Thus, it is likely that they worked because they had no other option.[55] "Widow" remained a category that overrode any specific occupation even when it was clear from other evidence that such women were following a trade.

CLASS FORMATION AND THE TRADE CYCLE, 1871–98

The trade cycle was central to class formation in late nineteenth-century Petrolia. The oil industry was a source of wealth for a handful of oil producers, merchants, and refiners, but it was also a source of considerable instability. Fluctuations in the income levels of the working and middle classes were central features of the class experience and characterized economic, political, and social relationships between the classes. These fluctuations affected the ability of Petrolia's middle-class men, and a handful of women, to act in the business and political world. They also affected the stability of class relationships between producers and refiners, and between employers and working men. A large number of small producers resulted in an extremely volatile and competitive industry. There was also considerable risk for producers engaged in an industry where the supply of crude was irregular and capricious. During the 1870s most petroleum refining was done closer to the markets along railway lines in towns and cities throughout southwestern Ontario, primarily in London, and in principal manufacturing and distribution centres such as Hamilton, Toronto, and Montreal. Until 1877 the structure of railway freights favoured the shipping of crude to London for refining. In 1868 the duty on imported illuminating oil was raised from ten to fifteen cents per wine gallon to offset a five-cent excise tax on domestically manufactured oils. Despite substantial trade protection in the form of tariff and non-tariff barriers, Ontario refiners were forced to contend with declining American prices, the relative ease of entry of imports, and an inferior product that limited access to overseas markets. Thus, the petroleum industry relied upon the refining of local crude oil for the tariff-protected domestic market.[56]

Several attempts were made by both producers and refiners to combine to adjust the prices of crude oil to serve their own interests in the trade. Price-fixing arrangements were quite common among Canadian businessmen, Michael Bliss argues; they denied the maxim that competition was the life of trade and instead advocated coming together to ensure one another "a living profit." The aim of the two major groups in the Canadian oil industry, John S. Ewing writes, was stabilization. Ewing explains: "To producers, stabilization meant a constant high price with a perpetual market for all the oil they could produce. To the refiners, it meant a constant price low enough to permit them to manufacture at a good profit; in times of declining consumption, this might mean a very low price indeed."[57] The price of crude oil was around $1 to $1.50 per barrel in 1870 and during the early months of 1871. In June 1871 the Lambton Crude Oil Partnership was signed by 105 oil producers from Petrolia, representing nearly one-third of all male householders in the town. According to the terms of the agreement all partners were required to dispose of their oil through the Board of Managers to whom the producers would furnish a statement of the quantity of oil they wished to sell. The Board of Managers would then sell the oil at the highest possible price and hand the proceeds to the producer, deducting a small percentage to pay for the expenses of operating the partnership.[58] By early July the price of crude was holding steady at $1.70 per barrel, a rate about equal to the cost of production. Sales were made mainly to export refiners from whom the Association had orders far in advance of their ability to fill them. In November the refiners organized the Oil Refiners' Joint Stock Company of Ontario and immediately reached an exclusive agreement with the producers' combination for the marketing of crude oil. The two groups agreed to base the price of Canadian crude on the New York price for refined. To the disgust of the producers, the price of crude dropped to between $1.20 and $1.25 per barrel during the fall and winter months.[59]

A meeting of Petrolia's oil producers was held in the Oil Exchange Hall on 18 April 1872 to address the demand of the refiners for a reduction of the price of crude oil. A motion for a reduction of the price of crude from $1.20 to $1.10 was rejected; the producers unanimously agreed that the price should remain at $1.20 per barrel. The members also decided to extend the partnership for another year after the agreement expired on May 20.[60] During the summer months the price of crude dropped to between $1.00 and $1.12 per barrel. Producers refused to sell at these rates and began tanking their oil. In October Pennsylvania oil producers shut their wells down for sixty days in order to raise the price for American crude. Canadian refiners shifted their entire refining capacity to the American market, resulting in a shortage of crude oil when the sudden demand proved greater than the supply. The following month the price of Canadian crude rose to $1.90. The discovery at Marthaville, three miles northwest of Petrolia, satisfied the demand for crude.[61]

Canadian oil producers lost the export market to the U.S. when a rich new field was discovered in McKean County, Pennsylvania, in the spring of 1873. Petrolia, which relied for its existence on the oil industry, was plunged into a depression that continued for the remainder of the decade. The Petrolia correspondent to the *Sarnia Observer* reported in April that "So discouraging is the prospect in Petrolia, that a great many labourers, as well as some skillful mechanics have left, or are about to leave Petrolia; the dull times having nearly put a stop to development." The following year the first of Petrolia's "foreign drillers" left for Java. Refineries throughout southwestern Ontario were shut down or else operated on short time, and their operations were limited strictly to the home market.[62]

Until the late 1870s London was the regional centre for refining, possessing the largest number of refineries in the province.[63] Occasionally London refiners would combine to depress the price of crude. In 1873 Petrolia oil producers countered by opening their own oil refinery, the Home Oil Works Company, which was organized as a joint stock company. The fifteen original stockholders comprised the community's leading oil producers. With the exception of James Miller Williams of Hamilton, all were residents of Petrolia. The producers expected to succeed by using their own crude oil. J.H. Fairbank was appointed President and Manager of the Company. The Town Council exempted the works from taxation for a period of four years in the interest of promoting the recovery of the local oil industry. The refinery covered an area of ten and one-half acres on the northern limit of the town and was built at a cost of $37,000. The Home Oil Works was a relatively large operation capable of refining 3,000 barrels of crude per week. They manufactured their own brand called "Prime White Oil," which, according to the Petrolia *Advertiser*, was unequalled for colour and sweetness by any other product manufactured in Canada. In the hands of the major oil producers, the Home Oil Works was a useful economic weapon in their intra-class struggles with refiners.[64]

Petrolia's oil producers were at the mercy of the refiners, while refiners benefitted from the increase in the retail price of refined oil and the decline in the price of crude. In the summer of 1874, Richard Herring, editor and proprietor of the *Advertiser*, endorsed the organization of another producers' combine. Herring wrote: "The producers can only be successful in an association that will resist the London refiners' influence." He explained:

> It will never do to let outside influences govern the Petroleum market either directly or indirectly. For a healthy protective association the great body of producers would co-operate, and no doubt a success would attend such a move. This would have the effect of regulating the oil market to the advantage of those who are legitimately entitled to a fair recompense for their investment, labor and risks.[65]

A week later Herring remarked that, "the agitation for a Producers' Protective Association has taken hold and is getting a favorable consideration."[66] On 2 November 1874 another producers' association, The Petrolia Crude Oil & Tanking Company, was formed. The company agreed to take all the oil from the producers at a fixed rate and sell it for the best price obtainable. At the end of the year the profits were divided in the proportion of 90 per cent to the producers and 10 per cent to the shareholders. Charles Jenkins was elected President of the company and Joseph McDougall was chosen Vice-President.[67] Pipelines were laid and receiving stations were built throughout the oil territory and an extensive business in piping and tanking crude oil evolved. The company also leased a refinery and began manufacturing its own oil for domestic consumption. Predictably, perhaps, given Petrolia's status as an oil-producing community, the editor of the *Advertiser* lashed out at Canadian refiners over the next several months. The high prices placed on Canadian refined oil allowed dealers to import American refined at lower prices despite the tariff. Consumers also preferred American oil to the foul-smelling Canadian illuminating oil.[68] The uncertainty of the oil market caused local operators and dealers to act with caution, and business in Petrolia remained depressed. In the spring of 1875, the *Advertiser* reported that several "prominent oil men" had departed for Manitoba to drill test wells for the federal government. Some oil producers attempted to make up for the decline in oil production by turning to farming. According to the *Advertiser*, "Almost every derrick and engine house had a piece of ground around them under cultivation."[69] The scarcity of money and the uncertainty of the oil market also resulted in high unemployment. "This state of affairs," the *Advertiser* reported on 25 June 1875, "is badly felt by teamsters, especially as horse feed is so high."[70]

The local oil industry rebounded temporarily in 1877. The price of crude oil hovered around $1.40 all year and there was a considerable amount of drilling activity. In its "Weekly Oil Report," for September 23, the *Advertiser* reported that 503 wells were pumping, 14 new wells were being drilled, 8 wells were being repaired, and daily production averaged 1,614 barrels.[71] In October 1877 another producers' association, the Mutual Oil Association, was organized, comprising 154 producers. Four managers were appointed to conduct the working operations of the entire crude oil business: Henry W. Lancey, who was appointed chairman; Jacob L. Englehart; John McDonald; and A.C. Edward. John H. Fairbank and D.M. Kennedy were appointed trustees. None of the men was remunerated for his services, although all had interests in the oil industry. Oil would be sold for the best price obtainable, with the producers receiving the proceeds of two-thirds less three cents per barrel for office expenses until the price exceeded $2 per barrel, when a few cents would be retained to form a reserve fund for the association. The producer was given a receipt for the remaining one-third, which could be disposed of as the owner saw fit. A few weeks later Herring chastised the handful of producers who refused to join the Mutual Association for "letting, it would seem, personal spleen interfere with business principles." He wrote: "Beware lest there should be one or two among the long list of producers mean enough to take advantage of enhanced prices should an opportunity occur, without sharing the benefit with their fellow producers, which every honest man will only be too glad to do."[72]

The price of crude climbed to $2.08 per barrel in January 1878. London refiners countered by dipping into their stores of tanked oil. The producers' association hindered the refiners very little as they simply passed along the price increase to consumers. The storage of crude oil by the producers had been mounting ever since the formation of the Mutual Association. By the spring of 1879, a surplus of

380,000 barrels had accumulated, yet no arrangement had been made for the disposal of surplus oil. Crude prices dropped to $1.70, a price that did not materially help the small producer. The *Monetary Times* explained, "For instance, one man produces 30 barrels per day, which he reports to the Association, his amount for sale will only be about one-third, or say ten barrels; the rest will be tanked. Thus, he receives $17, less expenses, whereas, if he sold it on the open market at say 80c per barrel, he would receive $24."[73]

Attempts to renew the producers' combination after the initial one-year term expired dragged on for several months. The oil trade was at a standstill and wells were shut down. Many of those who refused to sign the bond to form a new association were small producers who stated that they could not produce their oil at the price paid by the Mutual Association. A major obstacle that confronted the producers was the uneconomical way in which most wells were operated. Many were poorly cased and when the flow of oil declined water leaked in, resulting in an increase in the cost of pumping. On 1 May 1879, "Black Friday," the association collapsed as producers frantically sold their oil to keep their creditors at bay. In the aftermath of the demise of the producers' association, one small producer, in a letter published in both local newspapers, declared that "Sensible people are looking about them, and find that they are emerging from a transition state of helplessness into a state of self-dependence." "Supply and demand govern every business, and crude oil is no exception," the correspondent continued. "Over production is the bane of our industry and rings and corners have been instrumental in bringing about the present unhappy state of affairs."[74]

In the immediate aftermath of the collapse of the Mutual Oil Association the price of crude hovered between 50 and 55 cents a barrel. The larger producers discharged most of their workmen. Nevertheless, in a letter to the editor of the *Advertiser*, Charles Jenkins, President of the Petrolea Crude Oil & Tanking Company, wrote optimistically, "Fully five weeks have elapsed since the Mutual Oil Association here expired, and the market has been open for that time ... The first uncertainty and feeling that leads to panic has died away. The market has settled to a point, and crude oil exhibits an undertone of strength that shows bottom has been touched."[75] Throughout the summer months the large stock of crude, the absence of an export market, and the low price of U.S. oil combined to render impossible high-priced Canadian crude.

By the fall of 1879 movements in crude were "slightly more brisk than usual," and the market advanced slightly, holding firm at 67 cents a barrel. The *Topic* declared that, "the present depressed state of affairs cannot long continue, and now that petroleum is almost daily being utilized in some new branch of industry, thus rapidly increasing the consumption and demand, a 'boom' may at any time set in which will land our product triumphantly on a paying basis."[76] In the months following the dissolution of the Mutual Association the crude surplus declined by approximately 130,000 barrels. The town's refineries were running full blast. Herring remarked in the *Advertiser* "Petrolia is, therefore, receiving the benefit of having the crude manufactured at home – where it ought to be."[77] The improvement in the crude market was attributed to three main causes. First, the federal Conservative government passed the Petroleum Inspection Act on 15 May 1879, which introduced a differential "fire test." Canadian oils had to withstand heating to 105°F without "flashing," while imported petroleum was required to stand a flash test of 130°F. The legislation also discriminated largely in favour of the home trade in reference to fees for inspection. Canadian oils were charged ten cents per barrel and imported oils 30 cents per barrel. Second, the low price of refined oil stimulated consumption. Third, new uses for petroleum were being discovered, most notably for the manufacture of gasoline.[78]

In December 1879 Petrolia drillers organized an association in an attempt to secure their share of the profit from the rejuvenated oil industry. Members of the drillers' association were bound to accept contracts for wells at the following prices: $250 for drilling and furnishing the boiler, the engine, and the rig; $220 for providing the men, but not the equipment; $150 for providing the men only; and $125 for simply drilling the well. A correspondent who signed his letter "Engineer" wrote to the editor of the *Topic* disparaging the fact that no provision had been made for the scaffold men and engineers who worked in the oil fields. He suggested that the improvement in the oil trade should confer its benefits on the working men as well as the producers.[79] On 4 March 1880 a letter addressed "To Working Men" and signed "One Interested" appeared in the *Topic* encouraging manual labourers to organize. The correspondent wrote: "A short time ago there was a combination formed for the purpose of raising prices in drilling, and, so far, as can be ascertained, succeeded in doing so, but where is the profit to the men who do the manual labor, promised by that association, for the majority of them are still working for the paltry sum of ten and twelve dollars per well." A few weeks later, on April 22, another letter to the editor appeared in the *Topic* asking oil well labourers to organize in order to secure uniform wages, and announcing that a meeting would be held in Fairbank's Hall to form a union.[80] No further evidence was found, however, to suggest that any labourers' union existed in Petrolia prior to the organization of a Knights of Labor Assembly in 1885.

Persistent attempts to forge combinations among producers and refiners failed, in part, because of the inability to restrict the entry of new competitors. The formation of Imperial Oil in 1880 represents the first significant concentration of capital in the industry. Instead of relying on the payment of rentals to obtain the support of other refiners, as had been the case with previous refiners' combinations, Imperial Oil was based on the "Rockefeller Plan."[81] The Imperial Oil Company was organized as a joint-stock company, with a capitalized value of $500,000. Imperial Oil encompassed the ten leading refiners in Ontario, including Petrolia refiner Jacob Lewis Englehart, who owned the Silver Star Refinery, and most of the refiners formerly involved in the London Refining Company. Englehart became Vice-President of Imperial Oil a position he retained until his death in 1921, even after a controlling interest was purchased by Rockefeller's Standard Oil Trust in 1898.

In July 1880 the *Monetary Times* reported that "[t]he concentration of refining power by the Imperial Co., and their systematic attempts to absorb competing refineries," made a steady price for refined oil a certainty. Over the next several months the Imperial Company absorbed competing refineries and concentrated its operations in Petrolia.[82] By the fall of 1880 the price of crude oil had climbed to $2 per barrel; in October it peaked at $2.15 per barrel. The increase in the price of crude oil was attributed to a combination of the National Policy tariff, increased consumption of petroleum, a growing variety of uses for oil and its products, vast improvements in the quality of domestic oil, and the enactment of the Petroleum Inspection Act.

The refiners and producers in Petrolia showed some tendency towards co-operation rather than competition. Many of the members of Imperial Oil were also large producers who turned their crude into the Company for refining. The dominant perception among producers was that refiners were not dealing with them equitably. In November 1882, with stocks of crude at an estimated 300,000 barrels and the price of crude at $1.50 per barrel, the Producers' Oil Exchange Association was formed. The following year the Oil Exchange Financial Association of Petrolia was incorporated, and capitalized at $150,000 for buying, selling, and advancing money on petroleum and its products.

In its year-end summation of the state of the local oil industry, the *Advertiser* commented: "Crude this year has been depressed all through, while the refiners have combined, and have been and are getting the very best prices for their production."[83] The price of crude continued to slide and by mid-January 1884 was selling at between 70 and 75 cents per barrel. Producers began tanking their oil in an effort to force crude prices upwards. W.H. Hammond organized The Producers' Tanking Company for tanking, piping, and warehousing crude. During the summer months the price of crude remained at 75 cents per barrel and producers talked of shutting down operations for a couple of months. A public meeting attended by 75 producers was held on 22 September 1884 to devise a means to secure a better price for crude. Many of Petrolia's largest producers argued that to help the crude end of the business, producers had to once again secure an interest in the refining business. John D. Noble argued that building refineries would not elevate the price of crude oil. While he conceded that it would be better for producers also to be refiners, not all producers could be refiners. Instead, Noble suggested the refiners ought to be placed in the position of manufacturers, and the owners of the oil should make the money. Alluding to conditions in the U.S. oil industry, Noble suggested that a Petrolia stock exchange be established and that the four tanking companies – the Imperial, H.W. Lancey, the Producers Tanking Company, and the P.C.O. & T. Company – should come together and make their receipts uniform. According to Noble, only through the introduction of a system of brokerage would producers realize an increase in the price of crude oil. J.H. Fairbank endorsed Noble's idea of forming an exchange that would function as a board of trade. He further suggested that shutting down the wells would have the immediate effect of driving up the price of crude oil. Committees were organized to organize an oil exchange, prepare a refinery scheme, and investigate the shutting down of production on Sundays.[84]

The Petrolia Oil Exchange opened on 23 December 1884, with thirty-four major oil producers as the charter members.[85] J.H. Fairbank, Petrolia's largest oil producer, banker, and hardware merchant was elected president. A board of management was chosen comprising

Petrolia's largest producers, some of whom were also refiners, namely, Edwin D. Kerby, Martin J. Woodward, William H. Hammond, Jacob L. Englehart, Robert I. Bradley, James McCort, James Kerr, Alexander C. Edward, William J. Gibson, Robert Morris, Charles Jenkins, and John D. Noble. The Exchange was open daily for business from noon to 12:30 p.m., Petrolia west end bell time, with the exception of Sundays and holidays. Members were required to report all transactions on "'Change, whether as buyers or sellers, effected each day, to the Secretary before 3 o'clock p.m."[86] James Kerr, Secretary of the Petrolia Oil Exchange, informed the Royal Commission on the Relations of Labor and Capital on 14 January 1888 that, "The oil exchange is simply a market place where people who wish to buy or wish to sell, if they have the *entré* [sic], go, every day if they choose, either for themselves or for people who are not members of the exchange. It is simply a public market." When asked by Commissioner Armstrong what benefit the producers of oil obtained from the exchange Kerr responded: "There is this that accrues: there is a market every day where his oil can be sold for cash, and every day he knows the ruling price, without going around and making enquiries of various purchasers or other parties." Fluctuations in the price of oil as determined by the members of the Oil Exchange affected the entire community, and the residents of Petrolia paid close attention to the activities on the Exchange. The Petrolia Oil Exchange was one of the more enduring producers' associations, lasting until 1897, when U.S. oil took over the market.[87]

The public meeting of oil producers also resulted in the formation late in 1884 of the Producers' Oil Refining Company. The company was established with a capital stock of $50,000 and the stockholders of the company were composed of fully two-thirds of the producers of crude oil, totalling 130 men. Robert Menzie, the Manager of the Producers' Refinery, and an oil producer, stated that the aim of the company was "to shorten or make smaller the gap between refined and crude margins."[88] The president of the company, Tronson Draper, was one of the largest producers in the town, and the vice-president, Robert I. Bradley, was deputy-reeve and a manufacturer of nitroglycerine. As was the practice in many nineteenth-century small towns eager to promote manufacturing, the Petrolia Town Council exempted the Producers' Oil Refining Company from taxation for five years.

In October 1884 the committee appointed to visit the wells that were operating on Sundays reported that 193 wells were running full out. The committee on Sunday shut-down further reported that the men who were running the wells expressed the hope that the committee would be successful, as they were eager to have Sunday to themselves.[89] The Lord's Day Act prohibited all persons from exercising their worldly labour or the business of their calling on the Sabbath, except the conveying of travellers and Her Majesty's mails by land or water, and other works of necessity. Oil producer Thomas McKitrick told the Royal Commission on the Relations of Labor and Capital that the issue of necessary labour was at the heart of the controversy of Sunday operations. McKitrick explained, "There are some wells in which there is considerable water, and the companies or parties owning them say that it does not pay them to shut down on Saturday and start on Monday, because the water accumulates to such an extent, that they cannot make the wells pay unless they run them every day of the week, Sunday included." It was "necessary" for small producers in particular to operate on Sunday if they were to realize any profit from crude oil production.[90]

One small producer had his gas-burner and feeder pump regulated so that the machinery would run from Saturday night until Monday morning without any attention. Herring remarked in the *Advertiser*, "The man actually does no work, goes to church and listens to a minister preach against the desecration of the Sabbath and returns home possibly as piously inclined as the clergyman himself, and perhaps saying less against his neighbours." Herring further pondered whether it would be "a cardinal sin" to define the running of wells on Sunday as a necessity.[91] The Sunday shut-down committee had only limited success. Most producers continued to operate on Sunday, which only compounded the problem of over-production and low crude oil prices and the payment of considerable sums every year for tankage.

Tension between the refiners and the producers over the discrepancies in the market for crude and refined oil persisted in the late 1880s and 1890s. Small producers complained that the Petrolia Oil Exchange failed to advance their interests in the oil market. In a letter to the editor of the *Advertiser*, "An Old Producer" wrote: "Since the formation of the thing called 'The Petrolia Oil Exchange' producers of crude oil have more than ever been at the mercy of refiners." According to the producers, the Exchange was merely a kind of central office for managers of refineries to congregate and more readily determine the price of crude. The correspondent commented: "If refiners were only honest men, and a few of our leading producers were not so hoggish, things would now very soon right themselves – either by an honorable arrangement for the price of crude according to what refined sells for or ceasing the drill until our production more evenly matches the consumption."[92] Herring also suggested that the Exchange did not advance the interests of the producers. He agreed with the original idea of buying and selling all crude oil on the Exchange, so that if a producer came along with a thousand barrels of oil, competition would occur amongst the refiners and the highest price would be paid for the commodity. According to Herring, however, such had not been the case in the operation of the Petrolia Oil Exchange. "Most the large blocks of oil brought to the market by the producer never see the Exchange at all, and all the refiners produce more or less of their oil."[93]

Workers employed in the oil industry did not fare well during the 1880s. Menzie testified that men employed in the refinery worked only seven months of the year, beginning in May or June, when oil was refined. The shipping season extended from September through October, falling off from November onwards, and by March business was flat again. During the peak period from June to October, the Producers' Refinery employed as many as thirty men, and during the slow season from March to May as few as seven or eight. Coopers constituted the "movable" portion of the labour force. As Menzie explained, "They know the season as it comes along and they come here and go to work, and they go away again; some go to Cleveland and the American oil fields." According to Menzie, skilled workers were sometimes hard to come by as skilled labour for drilling was exported to other parts of the world.[94] When asked about the conditions of workers in the Petrolia oil industry, David Mills, an oil producer, stated, "I do not think they [working men] have got very fat in these last five or six years."[95]

In November 1885 an assembly of the Knights of Labor, a major working-class organization, was organized in Petrolia. Greg Kealey and Bryan Palmer have argued that the Knights of Labor in Ontario created, if only briefly, a distinct "movement culture," built on working-class values and ideals. The Knights were not only strong in urban centres, particularly Toronto and Hamilton, but also in small towns throughout Ontario, wherever a particular industrial activity predominated.[96] Not surprisingly, therefore, the Knights gained a stronghold in the oil-producing centre of Petrolia. Reliable Assembly 4570 existed for only two years, but had a large membership of 500. Lectures were held in Petrolia and featured notable Knights of Labor figures from both Canada and the U.S., including F.D. Phillips from Stratford, Daniel J. O'Donohue from Toronto, Dick Trevelick from Detroit, and Frank M. Fogg from Lansing, Michigan. The speakers focused on the aims and objectives of the Knights of Labor and addressed conditions in the oil industry. In his lecture Frank M. Fogg referred to the Standard Oil Company as a set of giant monopolists who "were sucking the lifeblood" from the Canadian oil trade.[97]

Co-operation was a fundamental principle of the Knights of Labor, and support for co-operation was undoubtedly found in Petrolia where Reliable Assembly attempted to organize a refinery, a flax manufactory, and a general store. The Recording Secretary notified the *Palladium of Labor* in August 1886 that the proposed oil refinery would "give a fair price to the producer for the crude, and place the oil business in a more satisfactory shape than it has ever been."[98] Ultimately, however, the cooperative initiatives of the Knights of Labor in Petrolia were unsuccessful.

In his 1890 testimony before the Royal Commission on the Mineral Resources of Ontario, J.H. Fairbank reported that nine-tenths of the oil of this country was manufactured in Petrolia, and that increased the number of men employed very considerably.[99] The oil manufactured locally was destined for the home market. Petrolia's producers and refiners were unable to compete with their U.S. competitors. The U.S. dominated the export market to Britain by producing better quality and cheaper refined than Canada. John D. Noble suggested that the principal reason Canadian oil manufacturers were unable to compete with the Americans in the English market was that they produced about 75 per cent illuminating oil compared to Canada's 40 per cent. Some U.S. wells produced as much as 2,000 or 3,000 barrels a days, while Petrolia wells averaged only half a barrel a day. Noble cautioned that "were the American oil allowed free into this country our industry would be completely wiped out; we could not compete with their flowing wells."[100]

The U.S. conglomerate Standard Oil was eager to infiltrate the Canadian market. Among the many companies that made up the Standard Oil combination, the one most directly concerned with Canada was Standard Oil Company (New York). By the late 1890s that company was directing a chain of subsidiaries that spanned Canada almost more completely than did Imperial Oil. The Canadian subsidiaries included the Eastern Oil Company in Saint John and Halifax; the Bushnell Company in Montreal, London, and Sarnia; the Queen City Company, whose primary holdings were in Toronto and Ottawa; and the British Columbia Oil Company, operating in Vancouver and other parts of British Columbia. The Petrolia-based Imperial Oil Company simply did not have the assets to compete with Standard Oil. The first negotiations for sale, initiated by the management of Imperial Oil beginning in 1895, dragged on for three years, with an unsuccessful outcome. In the interim it seemed as though Standard Oil would take over the Canadian oil industry.[101]

In 1896 the Bushnell Company purchased the Refinery of Fairbank, Rogers and Company in Petrolia, and the following spring bought the Alpha Refinery in Sarnia, which was no longer used. The Alpha Refinery was a large establishment with a capacity of 10,000 barrels of crude oil per month. The plan was to use the Alpha Refinery to supply the Standard subsidiaries in Canada with high quality products refined from Canadian crude. On 3 June 1897 the *Advertiser* reprinted an article from the *New York World* under the heading, "Story of the World's Greatest Fortune, The Crimes that have made the Standard Oil Company worth $325,000,000!" The *Advertiser* suggested that this was a narrative of "Lying, Perjury, Bribery, Arson, Conspiracy to Murder, Sanctimony," and that this institution was making "strenuous efforts to strangle the Canadian oil refiners and producers."[102]

Controlling interest in the Imperial Oil Company was sold by its shareholders to the Standard Oil Company in June 1898. The sale included Imperial's operations in Petrolia, the National Oil Company, and the Petrolia Crude Oil and Tanking Company. The *Advertiser* remarked, "This virtually embraces all the oil interests of Petrolia and gives the Standard complete control of the Canadian oil business."[103] Shortly thereafter the Petrolia refinery was torn down and Imperial Oil removed to Sarnia. The heyday of Petrolia as the centre of the nation's oil industry was over.

CHAPTER FOUR

A Respectable Victorian Town: Public Space, Voluntary Associations, and the Creation of a Culture of Refined Sociability

Petrolia oil producer, merchant, and banker, John Henry Fairbank typified the Victorian paternalist employer. His sense of Christian stewardship and his role as a guardian of social respectability and moral conduct earned him the sobriquet "Father of Petrolia." When his employees presented him with a Christmas gift in 1896, Fairbank's paternalist inclinations surfaced. In expressing his gratitude to his employees Fairbank stated: "We have worked together; while you have bent your backs to the oars, I traded to steer our little boat away from rocks and shoals." As the captain of what Fairbank described with humility as a small ship, he expected loyalty from his employees, which he viewed as having a "broader meaning than attachment to a prince or obedience to the law." "Within its scope are fidelity, constancy, devotion," Fairbank remarked.[1] Fairbank's vision of himself as the captain of a small ship crewed by loyal employees extended beyond his business enterprises, and incorporated his efforts to secure the respectability of the town by transforming Petrolia from a rough frontier mining community into a model of civic virtue.

For over fifty years, until his death on 10 February 1914 at the age of 83, Fairbank was a philanthropist, who used his personal fortune to build community institutions. He provided a public space, a hall located over his East End hardware store, for meetings, lectures, balls and commercial entertainments. Fairbank Hall also served as a meeting place for the village council. He was a member of the local Anglican Church. In the spring of 1866 the first Anglican service in Petrolia was held in the barroom of the Fletcher Hotel. Fairbank gave the church the free use of Fairbank Hall for six years until 1872 when a permanent building was constructed on land he donated. Over the course of his life he was a member or patron of numerous voluntary societies and sporting and social clubs, including the exclusive Petrolea Club, a social club of the town's leading businessmen formed in 1887, and the St Andrew's Society. Although he was born in the U.S., Fairbank became a British subject, and he served as Reeve of Petrolia for three years from 1868 through 1870. He was also Chairman of the Board of Health and Warden of the Fire Department for several years. In 1882 Fairbank, a Liberal, was elected to the House of Commons, where he served for four years. Fairbank's public service was summed in Beers's *Commemorative Record for the County of Lambton,* as "a life full of action," and "he justly enjoys a reputation for public spirit and enterprise beyond the ordinary."[2]

Fairbank's charitable influence upon town life was welcomed. Like other middle-class male philanthropists in late nineteenth-century Ontario, Fairbank's public involvement was seen as benefitting the town as a whole rather than his personal business interests. Public space in latter nineteenth-century Petrolia was defined primarily as the terrain of men. Middle-class men enjoyed a host of political, commercial, and social options in their pursuit of power in the public sphere. As Joy Parr writes in her study of gender and class formation in two Ontario towns: "Men's credibility in the community was not grounded in their maleness per se, not in their market allegiances alone, but rested as well in their demonstrated responsibility as family men." Masculine entitlement to public office was based on their status as married men and was informed by convictions about what a good husband and father should be; what Parr describes as a well-articulated notion of "social fathering."[3] By taking up leadership roles in the community, Fairbank took on the broader responsibility of town father.

On 1 September 1855 he married Edna Crysler, the daughter of Hermanus and Edna Crysler, United Empire Loyalists who settled in Niagara Falls. A year later their first son, Henry Addington, was born. A second son, Charles Oliver, was born in 1858. Fairbank went to Oil Springs in 1861, where he developed an interest in oil speculation while surveying property owned by Julia Macklem. Meanwhile, Edna managed the family farm in Niagara Falls for the next six years and cared for their two small children. The couple's finances were precarious and the family farm was heavily mortgaged to pay for Fairbank's ventures in oil production. Her father also mortgaged his farm to raise the necessary money to bail Fairbank out of debt. Fairbank's eventual success as an oil producer might be attributed to a considerable extent to the labour and financial resources that Edna Fairbank brought to the marriage.

Edna Fairbank joined her husband in Petrolia in 1866. A few months later she gave birth to another son, Frank Irving Fairbank, who died on 7 August 1867. A fourth son, Huron Hope Fairbank, born on 2 June 1868, died two months later. The couple's only surviving daughter, Mary (May) Edna, was born on 25 October 1869. Edna Fairbank gave birth to another daughter Ella Lenore on 6 December 1871, but the infant died the following summer on 3 August 1872. Plagued by ill health throughout much of her adult life, and grieving the loss of her small children, Edna Fairbank spent several months at a Cleveland health spa between September 1872 and May 1873. She spent the next year travelling throughout England and Ireland with her young daughter May. They returned to Petrolia in September 1874.[4]

In 1889 Fairbank began the construction of a new family home, befitting his status as the "Father of Petrolia." Fairbank chose not to build his home in Crescent Park, where the homes of many of Petrolia's upper middle-class families were located. Instead, the mansion, which was named "Sunnyside," was built on a block of land overlooking Bear Creek at the eastern end of the business section, next door to the large white frame house where the family had lived for a number of years. The main floor contained a reception room, den, living room, and library. Ed Phelps writes, "On the third floor one arrived at the acme of social distinction in Petrolia – the Fairbank ballroom, whose wooden floor felt as solid as the street because it was indeed underlain with bricks."[5]

Although her health remained precarious, the next twenty years of Edna Fairbank's life, until her death on 7 March 1896, were shaped by her middle-class gender identities as wife, mother, primary organizer of the household, community volunteer, and society matron. The concept of "separate spheres," which asserted the naturalness of assigning to men active and dominant roles in the public sphere of commerce and politics and to women the nurturing and submissive roles in the private sphere of the home, was a powerful prescriptive ideology during the industrial revolution of the nineteenth century. Separate spheres, as Mary Ryan, Catherine Hall, and Leonore Davidoff have explained, were also integral to the shaping of middle-class identity. The home was strongly associated with a particular middle-class ideal of feminine domesticity, and gender was central to the strategies by which middle-class formation occurred during the course of the nineteenth century.

Upon her death, the *Petrolia Advertiser* described Edna Fairbank as a "public benefactress." The *Advertiser* eulogized:

> But when the Grim Reaper claims for his own a prominent citizen, whose life has been devoted to charitable objects, whose financial position has been such that she could carry out her unselfish plans with liberality, when the poor and the needy have been objects of her especial care. [sic] It becomes little short of a public calamity. Such a citizen was Mrs. Fairbank. Those whom she has befriended, either by pecuniary assistance or influence are many, and we believe that the majority of them have sufficient gratitude to at least revere her memory and keep it green in their hearts. Her's [sic] was a true charity – generally done in secret and therefore the more acceptable.[6]

Edna Fairbank's charitable work and volunteerism, most of which were done anonymously in keeping with the strictures of middle-class feminine gentility, allowed her limited access to the public sphere. She helped to raise funds, or used her own money, for a number of community causes, including the organization of the town band and the construction of the local Anglican Church.[7]

In the aftermath of the incorporation of Petrolia as a town in 1873, the success of oil producers, merchants, and professionals in the economic and commercial realm translated into power as moral authorities in the public sphere. The "rough" masculine frontier culture of the 1860s built around the oil fields, the taverns, and gambling, gave way beginning in the early 1870s to more formal methods of association. During the latter decades of the nineteenth century Petrolia's middle-class men and women shaped their gendered class identities around dominant Victorian cultural ideals of moral reform and refined sociability. Public space came to be defined primarily as the terrain of middle-class men, with less space for women, although middle-class women were certainly found in public and wielded some measure of power.

As the oil economy of Petrolia was integrated more extensively into global trading patterns in the early 1870s, this material context fostered the ideology of separate spheres as the local middle class consolidated. The home was strongly associated with a particular ideal of feminine domesticity based on the family and associated with the private sphere. Over the course of the nineteenth century the domestic sphere of women became a benchmark of middle-class identity. The public sphere was constructed as the province of middle-class men on the basis of exclusions, notably those of gender, ethnicity, and religion. In latter nineteenth-century Ontario, the middle class "model of masculinity," was defined as the Anglo-Protestant businessman, father, sportsman, and public official who embodied the ideals of masculine self-improvement. Women remained within the familial framework, and when they did enter the public realm it was primarily to represent the private domestic values built around the family and to contribute to the building of a culture of refined sociability. The experience of class was different for men and women. The gendered process of class formation in latter nineteenth-century Petrolia took place in a variety of locations beyond the workplace and the family, including voluntary societies, civic celebrations, and sports, and in ways that involved both consent and coercion. In the domain of free sociability and voluntary association a public culture was forged in Petrolia.

Historians of class formation, fraternalism, and volunteerism have characterized the Victorian era as "the age of societies." As Davidoff and Hall write, "The public character of such societies made them open and visible to all, and indeed public accountability was one of their principles."[8] The two Petrolia newspapers, the *Advertiser* and the *Topic*, announced upcoming meetings and reported on the election of officers to the various societies, including fraternal orders and societies aimed at the cultural and educational needs of the middle classes. The newspapers also contain accounts of feasts celebrating the anniversaries of patron saints, fraternal funerals, annual lodge church services, fundraisers, and reports of picnics, balls, and other community events organized under the auspices of a particular voluntary society.

MALE ASSOCIATIONAL CULTURE AND CLASS IDENTITY IN LATER NINETEENTH-CENTURY PETROLIA

Voluntary societies thrived in nineteenth century Ontario villages, towns, and cities. In 1887 Petrolia had a Mechanic's Institute, three Masonic Lodges, a St Andrew's Society, a branch of the Independent Order of Good Templars, an Ancient Order of United Workmen

Lodge, a Royal Arcanum Council, an Orange Lodge, one Court of Foresters, two fire companies, and one hook and ladder company. The most pervasive forms of voluntary organizations in late nineteenth-century North America were the friendly societies. Some of Petrolia's lodges, like the Orange Order and the St Andrew's Society were based on ethnic ties or religious identity. The Catholic Church's denunciation of "secret societies" throughout much of the nineteenth century meant that most lodge members were Protestants. Some lodges, like the Independent Order of Good Templars, were organized around various temperance initiatives. By the end of the century, insurance orders, such as the Knights of the Maccabees, dominated the local fraternal scene. Women's auxiliaries under the auspices of the mostly male fraternal orders, such as the Lady Maccabees and the Rebekahs were also organized.

Fraternal relations were grounded in a model of kinship where lodge "brothers" occupied a shared masculine world. These all-male associations offered sociability, charitable relief in times of distress, and business and political connections. Ritual and symbolism were used to create solidarity and to address concerns about class, gender, and other kinds of differences. The fraternal rituals affirmed the values of respectable manhood – sobriety, thrift, piety, self-restraint, moral obligation – all consistent with an Anglo-Protestant middle-class cultural orientation and befitting the needs of capitalism in nineteenth-century Ontario. Nineteenth-century fraternalism, as Mary Ann Clawson has demonstrated, based itself on a principle of exclusion from which it derives its power.[9]

Fraternal lodges reinforced the dominant middle-class ideal of separate spheres. In a letter to the editor of the *Petrolia Advertiser,* signed "Woman's Rights," a Petrolia woman disparaged the gender inequalities of men's fraternal culture. She writes, "We are expected to be contented at home, and yet night after night while those who should be our protectors at home are galivanting [sic] half of the night with societies making high pretension." In her letter she reveals that these men transgressed many of the virtues that fraternalism supposedly espoused, most notably sobriety and self-restraint. The woman continued: "Those who are in the habit of guzzling down strong drink till, under its influence, many feel satisfied, but it must be disgusting to men of sense to observe part of the company 'licking up such fantastic capers before high heaven as would make angels weep,' and witnessing the 'reason riot' of persons supposed to be respectable members of the community." She concluded with an appeal to justice and protested against "secret societies that ostracise the 'weaker sex'" by drawing a line of distinction against women. She further recommended that the men "leave the midnight gatherings to the lower order or intellect."[10]

One of the town's three Masonic lodges was named "Washington" in recognition of the large population of Americans who came to Petrolia during the early days of the oil boom and became permanent citizens of the town. The 1921 commemorative golden jubilee history of Washington Lodge also suggests that the name was chosen to represent the "universality" of the Order: "the Masonic brotherhood was not national or political, but international, even universal in its scope."[11] The bonds of fraternalism were truly international for Petrolia oil drillers who journeyed to "foreign fields" in the late nineteenth and early twentieth centuries. Neil Sinclair left Petrolia to work in Italy in March 1880, and spent the next twenty years drilling in foreign fields. A founding member of Washington Lodge and the Bruce Chapter of the Royal Arch Masons of Petrolia, Sinclair visited the quarries in Jerusalem where the stone used to construct King Solomon's Temple was taken. He removed a stone out of this ancient quarry and had it sawed into several pieces. From the forests of Lebanon he cut olive wood which was also used in the construction of the temple. When he returned to Petrolia in December 1900, Sinclair presented each of the local lodges with a set of working tools symbolizing the religious and craft principles of Masonry that he had made from these materials. The *Petrolia Advertiser* remarked that, "Although Bro. Sinclair has been away for nearly twenty-one years he never forgot his brother Masons in Petrolia."[12] Another driller, Morley Evoy wrote to his fellow craftsman in Petrolia Masonic Lodge from Persia in the winter of 1919. In his letter Evoy wrote that while en route the drilling crew stopped in Manila, where he met some officials at the post office who were members of the craft and who invited him to attend a lodge meeting.[13]

Occasionally, the ritualized all-male sociability of the fraternal lodge was put on public display, most noticeably during ceremonies to lay cornerstones for new meeting places and during parades. On 28 June 1887 a special meeting of the Grand Lodge of Ancient, Free and Accepted Masons of Canada was held to lay the cornerstone for Petrolia's new Masonic temple. The Grand Lodge was opened by the Grand Master, Most Worshipful Brother Henry Robinson in the ante-room of the Masonic Hall in the east end of town. The brethren in their regalia and jewels then formed a line two deep and marched to the site of the new hall accompanied by the Oil Springs Band. The Grand Master received a ceremonial trowel and with it struck the cornerstone three times stating: "Well made, well proved, truly laid, true and trusty, and may this undertaking be completed by the craftsmen according to the Great Plan, in peace, harmony and brotherly love."[14] In Petrolia, as elsewhere in nineteenth-century small-town Ontario, the halls built by the Masons and the Oddfellows occupied a prominent place in the townscape.

On 21 June 1897 a committee of men from the local Sons of England organized a giant procession to mark Queen Victoria's Diamond Jubilee, involving an estimated 700 members of Petrolia's friendly societies accompanied by the town band. According to the Petrolia *Advertiser,* the procession represented Petrolia's transformation "from an American town, full of Yankee speculators," where "the Fourth of July was equal with Dominion Day," to its status as a member of the British Empire.[15] The membership of Petrolia's male friendly societies thus played an important role in transforming the identity of the community from a rough, shanty town of unscrupulous "Yankee" oil speculators into a respectable, Victorian town that endorsed the British imperial connection and an Anglo-Protestant Canadian national identity.

Friendly societies provided ample opportunities for mixed-gender socializing. Concerts, dances, tea meetings, dinners, and fundraisers were organized with the wives, daughters, and sisters of lodge members in attendance. Lodge brothers marked the anniversary of their order by marching in full regalia to church, where they joined the congregation and listened to a special anniversary service from a local Protestant minister.

Working men's participation in fraternal associations alongside middle-class men has been understood in the narratives of some historians as providing working men with a collective experience that encouraged the formation of a unified working-class consciousness. In late nineteenth-century Ontario, Lynne Marks argues, a shared masculine gender identity mediated divisions between the working class and the middle class and promoted cross-class mutualism.[16] Others have suggested that voluntary associations were primarily bases of community power led by middle-class men of property and standing in the community.[17] Bryan Palmer points out that simple, monolithic explanations of the working-class experience are insufficient. He recognizes that the collective experience of lodge life was essential to the articulation of the negotiated values of working-class mutuality, but he also acknowledges that middle-class masculine ideals of morality and class hierarchy also had an impact on the identity of late nineteenth-century working-class men. "The friendly society," Palmer writes, "was thus an important locale in a two-sided cultural context in which workers could be pulled in contradictory directions."[18]

In the latter decades of the nineteenth century there were divisions between the working men of Petrolia (many of whom joined the Reliable Assembly of the Knights of Labor during the 1880s) and the middle classes. The middle class, made up of those with economic power, was deeply divided between oil producers and oil refiners, who aligned themselves behind various combinations. Participation in friendly societies allowed a fractured and divided socio-economic group to act as a class, which was necessary if middle-class privilege and power were to be stabilized and extended. The fraternal order was an ideal way of responding to class differences while at the same time making possible a limited degree of class action and class expression.[19]

Unfortunately complete membership listings for Petrolia's lodges, which could be used to form a conception of the class structure of voluntary associations using occupational categories, are not available. The only such compilation found was for Washington Lodge No. 260 of the Masonic Order. Washington Lodge was organized on 18 September 1871. Between 1871 and 1899, 154 men joined the lodge; the occupations of ninety of these men could be determined using linkages to census manuscripts, tax assessment rolls, and business directories. Within the group identified by occupation, the majority, 64 (71.1 per cent), were from the lower middle class, including 12 retail merchants and shopkeepers, 12 white-collar workers, 33 small manufacturers, most associated with the oil industry, and 5 hotelkeepers and owners of livery stables. A substantial proportion of the members of Washington Lodge, 22 (24.4 per cent), were skilled working men, among them two oil drillers and 10 engineers, most of whom also worked in oil production. Also included among the skilled workers in the lodge were a blacksmith, a brakeman, a cooper, a butcher, a machinist, a painter, a printer, a police constable, and an undertaker. Only five of the members identified (3.2 per cent) were from the upper middle class.

Listings of voluntary society officials in the two Petrolia newspapers, the *Advertiser* and the *Topic*, are useful indicators of the class composition of the leadership. From 1871 to 1899 inclusive, 963 men and 56 women were identified as officers. The women belonged to temperance lodges and the Lady Macabees. The occupations of 627 of the men were determined. Within this group, 356 (56.8 per cent), were from the lower middle class, including 78 (12.4 per cent) merchants and 92 (14.7 per cent) white-collar workers. Most of the lower middle-class lodge officials, 173 (27.6 per cent), were master artisans or small manufacturers involved in the oil industry either as producers or refiners. The officers of Petrolia's voluntary societies were also drawn extensively from the ranks of the skilled working class, 174 (27.8 per cent). Thus, the majority of officers between 1871 and 1899 were from the lower middle class and the skilled working class, with only a small proportion, 81 (12.9 per cent), from the upper middle class.

Particular attention must be paid to how men were recruited to class places through voluntary societies. As Darryl Newbury points out in his study of the fraternal milieu in Ottawa-Hull during the nineteenth century, associational life was as much a repudiation of egalitarianism and collectivity as it was an affirmation of mutuality, with its confirmation of class hierarchies and individualism.[20] Fraternal hierarchies, either office holding or higher degrees, brought about the development of elite groups. Fraternal organizations, as some historians have argued, were the basis of the formation of a middle class from which men were recruited into public places. In nineteenth-century small-town Ontario, Christopher Anstead writes, lodges offered men, mostly from the lower middle classes of non-manual workers and skilled manual workers, experience in community leadership and the opportunity to influence the cultural identity of the community.[21]

In 1870 twenty-two-year-old Alfred T. Gurd, a farmer's son from neighbouring Moore Township, began in the oil business in Petrolia as an independent oil producer. By the early 1900s he was general manager of the Petrolia Torpedo Company and President of the Producers Oil Association. On 25 September 1879 he married Miss Dell Shaw. Five children were born to the couple: Dell, Kathleen, Jessie, Marian, and William. The couple belonged to the local Anglican Church.[22] Gurd was the first candidate initiated into Washington Lodge of the A.F. & A.M. on 2 January 1872. He advanced through the first three degrees of Masonry – Apprentice, Fellow Craft, and Master Mason – acquiring a knowledge of Masonry and its lessons of moral uplift and self-improvement. Membership in Washington Lodge provided Gurd with an avenue for attaining prestige, not only by rising through the various degrees but also through office holding. It was the general custom in Masonry that an officer "pass through the line," an automatic confirmation of a process that took place over several years.

Gurd was elected Secretary of Washington Lodge in 1873 and Worshipful Master four years later. He was Secretary of Washington Lodge again in 1881 and 1882, and in 1885 he became principal of a higher order of Masonry, Bruce Chapter. During the 1880s he held offices in the Royal Arcanum and the St Andrew's Society. Gurd's involvement in the associational life of Petrolia certainly provided him with leadership experience. During the 1890s he became more prominent in the public affairs of the town. He served as Mayor of Petrolia in 1890–91. A Conservative in politics, Gurd was elected to the provincial legislature in 1894 and held that position until 1899.

Like Gurd, several other middle-class men and skilled working men attained status in the community by holding multiple offices in Petrolia's voluntary associations. Besides helping men to meet and demonstrate the ideals of masculine respectability and to make business and political contacts in the community, office holding and membership in more than one friendly society may have been tied to the role of male breadwinner. Gurd joined and held office in the Royal Arcanum, which provided sickness insurance and death benefits for its members. For nineteenth-century North American families, as George and Herbert Emery and Lynne Marks point out, the primary cost of sickness was loss of the family head's earning power.[23] In an era when there were no state-funded health and family assistance programs, friendly societies were the largest providers of sickness insurance. Many of Petrolia's young middle-class and skilled working men probably joined societies like the Royal Arcanum, the Knights of Pythias, the Oddfellows, and the Knights of the Macabees for the insurance provisions. Male participation in friendly societies, therefore, was not entirely separated from the private sphere of family.

The fire brigade was another popular all-male associational space in nineteenth-century small-town Ontario. In Petrolia, where the prevalence of the oil industry meant that the potential for fire and loss of livelihood always loomed large, membership in the fire brigade appealed to young men seeking the esteem of the community, male camaraderie, adventure, and manly competition. On 3 August 1867 a large fire occurred at the well owned by Captain King. The fire extended to the adjacent property owned by John D. Noble, burning derricks, storage tanks, refineries, and some 7,500 barrels of oil. In all twelve acres were ravaged by the fire, and the fire continued for two weeks until the earth ran into the well and choked it.[24] Some preliminary discussion concerning the organization of a hook and ladder company had taken place among the members of the village council in the months before the fire. It was not until more than a year after the fire, in the spring of 1869, that the council passed a by-law authorizing the organization of a hook and ladder company. Oliver W. Chamberlain, an oil operator, was appointed fire chief. He served until 1871 when Walter Oliver replaced him. William Hartery, a local wagonmaker, built a hook and ladder wagon at a cost of $200. The wagon was stored in the old village hall on Pearl Street at the east end of the village. Another serious fire occurred on 30 July 1872 when the "Big Still" owned by Parsons and Co. exploded and 2,000 barrels of oil went up in flames. In June 1873 the council passed a by-law for the organization of a steam fire engine company. The Andes Fire Company, named after a fire insurance company, was formed. On 18 August 1873 Council voted to purchase a hand or brake fire engine for $800 and authorized the formation of a fire company to operate the new engine. The company was named the Union Reliable Company, after an oil well by that name. Following the incorporation of Petrolia as a town in 1874, J.H. Fairbank became fire warden, a position he held until 1889. The town's three fire brigades received some funds from the municipal council and were responsible to the council.[25]

A "friendly" rivalry commenced among the three fire companies to get first water on a fire. Each of the three fire companies was a self-contained, disciplined unit that shared many of the features of fraternal orders. The engine house was an all-male space where members spent hours polishing and maintaining the equipment and socializing. Each company elected its own officers, charged dues, raised funds, and selected uniforms. Their skills were honed by regular drilling under the direction of fire warden Fairbank.

Petrolia's fire companies tested their skills against other volunteer fire companies at tournaments held in communities throughout Ontario. Fire companies, as Lynne Marks writes, also served as a vehicle of community pride and boosterism. The *Topic* reported that the firemen "brought back a good account of themselves from the demonstration given by their Watford brethren" on the 24 May 1879 holiday; "the Union Reliable division of the brigade winning the hose and cart race from seven competitors." A large crowd gathered at the Western railway depot to welcome the brigade on their arrival home. The *Topic* remarked that "the deafening cheers which greeted their appearance must be regarded as a flattering evidence of the appreciative interest taken in them by our citizens."[26]

Of the sixty-five officers of fire companies identified by occupation between 1878 and 1886, 34 (52.3 per cent) were lower middle class, 23 (35.4 per cent) were skilled tradesmen, 4 (6.2 per cent) were upper middle class, and 4 (6.2 per cent) were unskilled working men. There was a substantial representation of lower middle class and skilled tradesmen in Petrolia's fire companies, including several oil drillers and refiners. According to the fire log for 1884 to 1894, the majority of the fires in Petrolia occurred at oil refineries or on the drilling rigs.[27] The fire brigades could override class barriers at times, as well as cultivate notions of class difference through a hierarchy of office holding.

The Dominion Day celebration organized by the town's fire brigade in 1879 demonstrated the status of the fire brigade in Petrolia, and its role in encouraging community pride. A parade, athletic games, and an elaborate picnic prepared by the women of the community were organized to celebrate the holiday. Early that morning the Petrolia firemen met the Watford fire brigade at the railway depot. Upon disembarking, the Watford firemen and their Petrolia colleagues formed a square. From the "regions of darkness" King Cetewayo and his Zulu band emanated. "The high-toned King," the *Advertiser* reported, "was seated on a delicate Phaeton drawn by four young warriors and surrounded by his body guard." As Susan Davis writes, the Callithumpian parade suggests that the social order and the hierarchy of associational life in the community, where the middle classes tended to occupy positions of rank and title, could be challenged.[28] The "sons of Zulu" symbolized resistance to European colonization and the privileged class status of oil producers and firemen in the community. In his address to the crowd King Cetewayo stated: "Other lands have the olive and the vine – we put oil in everything. Other lands have gold and silver which may be stolen – our wealth is unbounded expectations of which no one can rob us." King Cetewayo ridiculed the esteemed status accorded the mostly middle class and skilled tradesmen who constituted the fire brigades, and pointed to their tendency to engage in behaviour usually associated with rough male masculine culture, notably drinking. He continued:

> Look upon our mighty men. Who shall see them and not tremble? Look upon our men of fire – yes, upon our firemen! When did the lunch basket call them to feed, or the officer say "let us drink," or the fiddle sound for the dance, or the command "go forth and put on your good clothes" be given, and find

them hesitant to obey? When? Say! Echo answers when? And who is able to teach our fire chief how to hang hooks and ladders and tin pails upon our walls? Who shall say to our Andes Captain "smoke," and have him refuse.

King Cetewayo concluded his address by mocking the representation of Petrolia as a community of respectable middle-class families. "Are not our Kerbys, Smiths, Lindsays, Gurds, Engleharts and Greasy Charley, the joy of all the women of the country?" the King asked. "Are not all our men *giants*? And are not our women *queens*," he continued. "Yes, ladies and gentlemen, our town is noted throughout the length and breadth of the land for having produced more old bachelors, old maids, young ones and crude oil, than any town in the world."[29]

While participation in the fire brigade conferred a measure of status in the public sphere on middle-class and working-class men, the culture of Petrolia's firemen was a mixture of rough and respectable ideals of masculinity. This was suggested in a poem entitled "What would the Public do without us," composed by a "Firemen's Friend," and published in the *Advertiser* on 9 April 1886. In the first stanza the poet acknowledges that there is a "certain class of men" who stigmatize firemen as rowdies. The poet defends the firemen, however, by pointing out that there are also rowdies in "good society." He writes that firemen "strive to act with propriety." The virtues of respectable manhood demonstrated by the town's firemen, including bravery, attention to duty, and steadiness, are highlighted in the poem:

Who with the fireman can compare?
For action he is every ready!
He's willing both to do and dare,
And does his duty brave and steady;
While growlers on soft couches lie,
And little seem to care about us,
We rush to conquer or to die,
What would such growlers do without us?

The protection of property, "Firemen's Friend" asserts, is the single aim of the firemen, and he concludes rhetorically, "What would the public do without firemen."[30]

Women played a role in conferring rough or respectable masculine status upon the firemen. Respectability was bestowed upon the fire brigade in 1875, when Mrs Edna Fairbank entered the annual Firemen's Ball on the arm of her husband J.H. Fairbank, who was in full regalia. The couple then led the Grand March and formed the set for the first dance.[31] Later, in November 1878, the Petrolia *Advertiser* pronounced the ball organized by the Andes Fire Company "both a financial and social success," although with only one or two exceptions the members of the company presented themselves for admission, unaccompanied by their partners. The *Advertiser* remarked, "the fact is undeniable that until they can induce their ladies to lay aside this unnecessary and disagreeable pride of position, and to grace these events with their presence at least, the Firemen's Ball will always lack its principal element of success."[32] The associational culture of the all-male fire brigade was not consistently viewed as a suitable place for respectable women in public.

Like the fire brigade, the town band, organized on 15 August 1869, was an all-male voluntary association and community institution. The band performed on civic holidays, at balls, at church picnics and socials, and at private house parties. The band serenaded the town's elite on their lawns during the summer, and played at the roller rink and at ice-skating carnivals. By the spring of 1879, however, it seemed that the Petrolia band was dead, owing to a lack of financial support to procure new instruments and an instructor. George Denham, a druggist, organized a new band – a brass band. He engaged Professor D. Roblins, bandmaster of the 26th Battalion of Middlesex, as instructor. Frank Smith donated a building near his oil refinery for band practice. A subscription fund coordinated by Mrs Denham raised $105.50 to cover the cost of 16 instruments purchased from Nordheimer's in Toronto. Over the summer, the ladies of the community were visible in public, raising funds for the new band. In August they organized a promenade concert, the proceeds of which were donated to the band.[33]

Meanwhile, the Petrolia Band, also known as "the Old Band," was reorganized in the spring of 1879 with twenty men. Thomas Symington was hired as instructor. In September, several young, single, middle-class women of the community, under the direction of Misses Cooley, Lancey, and Gordon, organized a benefit entertainment for "the Old Band." The town's leading citizens and their offspring performed and provided their patronage, including J.L. Englehart, Emma Lancey, J.J. VanTuyl, and Ralph Gillespie. A total of sixty dollars was raised to assist the band.[34]

The population of Petrolia was too small to sustain two bands and on 24 June 1880 they amalgamated.[35] By the early 1890s the band was languishing once again. In a 10 March 1891 letter published in the *Topic*, one correspondent, "Citizen," cited a lack of financial support from the town council as the reason for the lethargy. According to "Citizen," the bandsmen purchased their own instruments. While there were numerous capable musicians in the community, most of the town's working-class men could not afford to purchase instruments. This also detracted from the calibre of the band, the correspondent suggested, since the leader had no choice in the selection of talent when the members purchased their own instrument. In concluding, "Citizen" recommended that the council and the citizens join together and purchase a good set of instruments and "give the band boys the support and encouragement they deserve."[36] Subsequently, on November 10, Edna Fairbank convened a meeting at the family's home to organize a brass and reed band. Here the private, feminine sphere and the public, masculine sphere converged. The outcome of the meeting was the appointment of a committee made up of the town's business and professional elite "to organize, control, and manipulate" the newly organized Petrolia Brass and Reed Band. The members of the committee were: J.L. Englehart, B.S. VanTuyl, Harrison Corey, Dr Dunfield, Dr Sturgeon, Robert I. Bradley, G.S. Pitkin, Robert Herring, and E.J. Lovelance. J.H. Fairbank was elected honorary member. The *Advertiser* reported that Edna Fairbank had already raised $150 in subscriptions to start the band.[37]

Employment was found for the bandleader, Mr Beauchamp, as market clerk, and for a few other skilled tradesmen. The municipal council pledged a grant of $100. In February 1892, the *Advertiser* issued a call for more bandsmen, indicating specifically that positions could be found for tailors. By May 1894, however, the band had "practically disbanded," after the council tacked the job of market clerk onto the duties of Constable McRitchie, thus forcing bandleader Beauchamp to find employment outside of Petrolia. Several bandsmen, loyal to Beauchamp, immediately announced their intention to resign and handed over their instruments to Mayor W.G. Fraser.[38] The town band survived, although it remained on unsteady ground. In a letter to the editor, one correspondent lamented the decline of the town band. He wrote: "The mission of a Band in a community is her good. Perhaps no other single feature can contribute so much to a town's life and pleasure." The correspondent also pointed out the usefulness of the band to the citizens of the town. It performed at celebrations of all kinds, on holidays, and at public occasions of both joy and sorrow. The letter continued: "The mission may truly be said to express in musical form the voice of the people."[39] In April 1897 the council pledged a grant of $10 for each monthly concert played, in order to sustain the Petrolia Brass Band.

Like the fire brigade and the fraternal orders, the voluntary militia was another exclusively male bastion. Militiamen, mostly British immigrants or British-Canadians, considered their attachment to England to be supreme and unwavering. As Mark Moss writes, "The so-called Anglo-Saxon mission to civilize and extend Britain's influence had an enormous impact on the Canadian psyche." The Canadian militia ideal drew on the cult of the Loyalist, and was tied closely to support for the British Empire. In late nineteenth-century Ontario "manliness" was also constructed in militaristic terms, with the "warrior" becoming the quintessential symbol of masculinity. Courage, physical strength, and military discipline became paramount ideals of Victorian manliness.[40] Involvement in the militia was a sign of social status, or a means to elevation in social status. Militia service was largely confined to the middle classes, and provided urban and small-town men with an opportunity to participate more fully in the civic life of the community. Militia units were "political machines," and positions in the officer corps, an avenue of upward social mobility, were allocated with partisan patronage.[41]

Following the British-style regimental system, each local unit was territorially based. This fostered a strong sense of shared identity between the unit and the local community. The Petrolia Company, which was part of the 27th Battalion encompassing Lambton County, played an important role in public spectacle. They participated in ceremonial parades marking civic holidays and Dominion Day celebrations. The militia put itself on display during their annual two-week summer training. Drills, mock battles, and troop marches were staged. "Even in towns where the militia companies were popular," Lynne Marks writes, "the patriotism of the respectable community was tempered by a hostility to the rough culture of the annual military camps." Excessive drinking and rowdiness challenged the behaviour that small-town Christian folk held in high esteem.[42] Even so, Moss and O'Brien suggest, these public displays went a long way towards glamorizing the military in the public imagination. Visiting the militia camp was a favourite summer activity for the local population, and "lady visitors," in particular, were made welcome.[43]

In September 1879 the 27th Battalion made their first visit to Petrolia for annual drill, setting up camp on Dawson's farm about one mile south of town, on the banks of the Sydenham River. The men spent the first day setting up camp. The following day, a reporter from the *Topic* visited the camp and found the men engaged in drill. "The men, although apparently not very well up in their drill, are, on the whole, rather a fine looking body, the greater majority of them being farmers and farmer's sons," the *Topic* remarked. "A few days of camp life and camp work will make a wonderful improvement in their appearance on parade, it only requiring a little efficient drill to make them fine soldiers."[44] After a few days of drill considerable improvement was reported at the "march out" held through the town of Petrolia. The men reportedly "presented a very neat and soldier-like appearance, which was much admired by numbers of citizens who congregated on the streets to witness the march." On Friday evening a handful of young, single ladies from Petrolia took possession of the marquee and arranged a supper for the officers. A number of "tried and true bachelors" were detailed by the Colonel to the regiment to attend to the ladies, the event providing an opportunity for courting. The field movements, conducted on Saturday, attracted a crowd of spectators from town. On Sunday morning and evening, the volunteers distributed themselves among the various churches in the community and attended services. A church parade was held in the afternoon, again attracting a large number of civilians. The battalion marched from their quarters to the grove to the accompaniment of the band playing "Hold the Fort." The Methodist minister, The Rev. Cobb, and the Presbyterian minister, The Rev. John McRobie, conducted the service. The camp disbanded the following morning. Commanding Officer Colonel Campbell praised the men for their "orderly and soldier-like conduct;" there were no published reports of drunkenness or rowdiness among the soldiers in camp.[45]

"INDIFFERENT TO EDUCATION, VOID OF INTELLECTUAL TASTES, WEAK IN MORAL PRINCIPLE": THE MAKING OF A LITERARY CULTURE IN NINETEENTH-CENTURY PETROLIA

The building of a literary culture was another component in the transformation of Petrolia from a rough, frontier village into a respectable Victorian town, and allowed women to carve a public space for themselves in the community. A mixed-sex literary society existed in Petrolia from the early 1870s, and endured until the end of the century when other forms of literary entertainment took hold. As Heather Murray points out in her important path-breaking study of literary societies in nineteenth-century Ontario, "literary" study did not mean reading only, but also included the rhetorical arts of composition, memory work, recitation, debate, and elocution. The emphasis on the rhetorical arts was indicated in the designation of the society as the "Petrolia Literary and Musical Society." Literary societies, Murray writes, combined a wide range of cultural activities, such as musical interludes, current-events discussions, local history research, and social picnics intended to spark courtship among the young unmarried members. They were also a prime mechanism for self and mutual improvement in the settler societies of nineteenth-century Canada.[46]

The members of the Petrolia Literary and Musical Society organized the literary program on a rotating basis. Regular Friday night meetings of the literary society comprised routine business followed by a program of recitations, solos, readings, and discussions, concluding with the singing of the national anthem. The 18 April 1879 meeting consisted of a series of readings by lawyer George Moncrieff and oil producer Alfred Gurd, and musical solos by Miss Huggard and Mr McSween. An original charade entitled "Petroleum" was presented for the first time, no doubt using symbolism familiar to local citizens. Oil refiner Frank Kitteridge gave a presentation on the celebrated Jolly Wax Works, highlighting the peculiarities of the figures on display in the museum while they lived and the circumstances surrounding their deaths.[47] Subsequent meetings included Miss Langton's presentation of a piece of local history, sketching the organization of the society and its results; Mrs Kitteridge's elocution from Longfellow; and Dr Dunfield's recitation of a course to follow to secure happiness. The newspaper accounts of the society's programs substantiate the research in history-of-the-book studies that suggests that for nineteenth-century Ontarians reading was a group and public activity, rather than an individual and private experience.

Membership in the Petrolia Literary and Musical Society allowed women seeking a higher education to hone their writing and public speaking skills and to master the rules of order and parliamentary procedure. Mixed-sex societies, Murray argues, also appealed to older or married women eager to gain a foothold in the cultural and civic affairs of the town.[48] Two women, Mrs Frank Kitteridge and Mrs Alexander McDonald, served on the executive of the literary society in the capacity of vice-president. In late nineteenth-century Petrolia, the literary society was a combined social, cultural, and political organization. It allowed women to play a role in civic improvement through the organization of a mechanics' institute in Petrolia in 1879.

When the time came for George Moncreiff to deliver his valedictory presidential address to the literary society in April 1879 he chose as his topic, "The ways and means of forming a Mechanics Institute and Library in connection with the Literary Society."[49] On 24 April 1879 the *Petrolea Topic* called for the organization of a mechanics' institute, pointing out the role that the institution might play in nurturing the moral and intellectual development of the young men of the town. The *Topic* suggested that both sons and daughters might benefit: "Mutual improvement societies tend to expand the elementary education they receive at the public schools and by bringing them in contact with superior intellect, strengthens them for their combat with the world, and should receive the encouraging support of all." In the same issue, a letter to the editor from "A Young Citizen" referred specifically to the benefits a mechanics' institute would have for the young working-class men of the town. The correspondent identified the lack of organized leisure activities promoting self- and mutual improvement specifically as a problem for young men. He asked, "Cannot something be done to assist young men to a better sort of life than the aimless existance [sic] they lead here – something to prevent nightly crowded billiard rooms and impassable street corners? Our young men are not more viciously inclined than those of other places, but the absence of morally intellectual resorts makes it more conspicuous, and it is much to be regretted that such is the case." He concluded his letter by suggesting that the Literary Society initiate the organization of a mechanics' institute, noting that since "it embraces large representation of the wealth and intellect of the town, it certainly could not fall into better hands." Victorian-era social reformers emphasized the self-improvement of the individual. They believed that the mechanics' institute would elevate the skills of artisans and skilled workers through programs devised to impart scientific principles underlying manufacturing while also providing the lessons of moral uplift necessary for its members to be compliant citizens. These cultural goals could be achieved through the provision of lectures, libraries, reading rooms, and evening classes.[50]

Another letter to the editor of the *Topic* revealed that "oil matters and money matters generally," were unsatisfactory, and prevented the organization of a mechanics' institute and mutual improvement society.[51] The women of the Petrolia Literary and Musical Society organized a fundraiser for the proposed institute. Their efforts, however, met with some opposition from men in the community who objected to women's participation in public affairs. On 18 June 1879 "Success" wrote:

> I approve very much of the generous proposition and effort of the ladies of Petrolea in getting up a concert for the benefit of a Mechanics' Institute, but Sir, in the first place, I would like to see the young men and the Mechanics take the lead in this matter, as it will take a considerable amount of money to provide proper rooms and suitable furniture for a society such as a Mechanics' Institute. I would rather see men subscribe a dollar or two each than to be led by the ladies to a fifteen cent concert, especially as it is not leap year.[52]

The letter prompted a series of responses criticizing "Success" for his "ungentlemanly" conduct. One correspondent wrote, "The author is evidently one of those who think the ladies should take a back seat in everything. If the ladies of Petrolea are interested enough in young gentlemen to wish to provide a place of resort for them, where they may combine pleasure and profit, it would certainly be more becoming for "Success" and others who feel and think like him, to help on the work, and not designate the efforts of the ladies to accomplish that end as a fifteen cent concert." Another correspondent, "PHILOS," praised the efforts of the ladies. He wrote, "A great many of the ladies of this town are the wives of mechanics, and why should they not, if it's in their power to help their husbands or brothers in such a good cause, do it?"[53]

The initial attempts by the committee of the literary society to organize a mechanics' institute soon stalled. A mechanics' institute was incorporated in Petrolia late in 1879. Later, on 4 June 1880, a slate of officers was elected. Stephen S. Griffiths, a solicitor, was chosen president. James Kerr, a builder, was elected vice-president, and oil producer Henry Rosenburg became treasurer. The nineteenth-century middle classes, Robert Morris argues, used the voluntary society to transfer resources to another social group, or to sponsor through approval, support, or material aid.[54] Some evidence suggests that this was also the case in Petrolia. Members of the Petrolia literary society served on the executive of the mechanics' institute. James Kerr, vice-president of the mechanics' institute, was also president of the literary society in 1880. Henry Rosenburg, John Fraser, and James McSween served on the executive of the mechanics' institute during the early 1880s, and they belonged to the literary society. Thus, while the mechanics' institutes are generally understood as providing artisans and mechanics with access to educational opportunities, they also helped their middle-class promoters, who set themselves up as sponsors and used the mechanics' institute to further their interests.

At the organizational meeting of the Petrolia Mechanics' Institute, members decided to open a reading room. The *Topic* cautioned that attention must be paid in the selection of books and periodicals to meet the mental wants of such a varied community. According to the *Topic*, it would not be difficult to fulfill the intellectual desires of the earnest student, who "finds beauty in everything which appeals to his intelligence, and is satisfied with any species of literature which furnished food for thought." More careful attention, however, was required in selecting works that would entice the mechanic and the working man, who by virtue of their class status were "indifferent to education, void of intellectual tastes, weak in moral principle. Possessing no particular taste for reading, something must be introduced to excite their imagination and create a desire for further knowledge." According to the *Topic*, "light literature" – works of fiction – "innocent and attractive," and not "trashy productions which tend to disorder the imagination and lead it into channels of thought antagonistic to morality," should be amply provided to benefit the young mechanics of the town.[55]

John A. Fairbank provided rent free the premises for a reading room and lending library, which was opened to the public in August 1880. The literary society soon amalgamated with the mechanics' institute. In its report to the Minister of Education in March 1881, the Petrolia Mechanics' Institute listed 140 members. For much of the decade the Petrolia Mechanics' Institute functioned primarily as a lending library. Books were purchased from a $400 provincial government grant matched by funds raised locally.[56] Books in science and art as well as works of fiction, biography, history, travel, and poetry were acquired for the library. Subscriptions were taken out for a series of periodicals and newspapers including the *Oil City Derrick*, which contained a comprehensive overview of conditions in the U.S. oil industry. The reading room was stocked with the two local newspapers, the *Advertiser* and the *Topic*, which were probably provided free of charge.[57] The reading room was open every evening to the public, free of charge, and young men and mechanics, in particular, were encouraged to make use of it.

A subscription to the Petrolia Mechanics' Institute library was one dollar per year. A partial list of subscribers for the period from 1881 to 1889 contains a total of 352 names, of which 193 could be identified by occupation using the decennial census manuscripts, business directories and tax assessment roles. Class was then determined on the basis of occupation. With the exception of two women who had an independent income, the occupation of husbands or fathers was used to designate class status for women subscribers. In the case of one male student, the occupation of his father was used.[58] Within this group of 193 subscribers, 182 males and 11 females, 122 (63.2 per cent) were middle class, 26 (13.5 per cent) working class, 9 (4.6 per cent) unskilled workers, and 36 (18.6 per cent) middle class. While some of Petrolia's mechanics and artisans participated, the mechanics' institute did a better job of catering to the interests of the middle-classes in the community.[59]

In the mid-1880s men and women enrolled in evening classes under the art education program instituted by the Minister of Education in 1881, in co-operation with the Ontario School of Art and the mechanics' institutes. In 1888 the course consisted of twenty-five lessons of two hours each. Lessons were given in freehand drawing from flat examples, practical geometry, linear perspective, model drawing, and memory and blackboard drawing. The strict adherence to rule, line and form, as well as geometric design, had applications in the production of many manufactured goods. As historian Foster Vernon remarked, "there is no doubt that the art education program was a great success in all ways except perhaps in developing creative artists."[60]

By 1889 the Petrolia Mechanics' Institute was in a state of decline. The lending library languished and failed to attract new subscribers despite an advertising campaign conducted by librarian Charles Howell. In December 1889 the library committee recommended that the lending library be closed.[61] The Petrolia Mechanics' Institute never entirely succeeded as a mutual instruction society where mechanics and artisans learned the lessons of self and mutual improvement. Among Petrolia's middle-class men, the male debating society continued to be popular into the late 1880s, and provided a place for young men to practise the skills of public debate and learn gentlemanly etiquette. The Dufferin Club, named after the Governor General of Canada from 1872–1878, met every Wednesday evening. Over the course of the year

1887–88, the middle-class male membership resolved that country life was superior to town life, Napoleon Bonaparte was a greater general than Alexander the Great, it was better to be born poor than rich, capital punishment should be abolished, ambition is a vice rather than a virtue, war is in no case justifiable, single life is preferable to married life, and wealth has done more to produce crime than poverty.[62]

The members of the Petrolea Club, organized in December 1887, were the leading businessmen of the town. Jacob L. Englehart was chosen president and Dr Charles O. Fairbank was elected vice-president. The club contributed to the building of a literary culture in the town and provided a separate male-only associational space for the elite of the business and professional sectors. Builder James Kerr, who was probably a member, wrote that its objective was "solely to furnish healthy and agreeable diversion to its members and guests." The club occupied a suite of four rooms in the Archer block, with reading, smoking, and amusement rooms. The reading room was abundantly supplied with English and American reviews, magazines, and illustrated journals, as well as the leading Canadian daily newspapers. The amusement room was outfitted with billiard tables, and the smoking room was furnished with small tables for chess, draughts, dominoes, and other board games. The discussion of politics and religion was strictly prohibited, as were profanity, gaming, and tippling. Kerr remarked, "the most rigid disciplinarian could not be more strict as to general deportment than is expected by the rules of the club."[63]

THE CREATION OF A CULTURE OF REFINED SOCIABILITY: ASSEMBLIES, CIVIC CEREMONIES, AND SPORT

In the 8 December 1924 issue of the *Saturday Evening Post*, Canadian writer William Alexander Fraser's short story set in the fictional town of "Petroville" was published. A couple of weeks later, the story was re-printed in two instalments in the *Advertiser-Topic*.[64] Fraser, who established his early reputation as a writer of animal stories, lived in Georgetown, Ontario. He was a frequent visitor to Petrolia at the peak of the oil boom. As the *Advertiser-Topic* pointed out, Fraser's story was obviously based on Petrolia; the names of the characters were easily recognized as citizens of the town.

Fraser's Petroville was a "joyous town; men laughed, and fought, and laughed again; their blood ran rapid, titillating to the exhilarating passion of the gamble that was in oil." Speculation in oil resources nurtured a rough culture built around gambling. Respectable leisure amusements were almost non-existent in Petroville. "There were no theatres, no movies in those days; there was not even a hockey, a lacrosse or a football team." Foot-racing was the passion of Petroville's male residents, and betting on races was widespread. "They'll bet on most anything, but twice on a foot race." The plot of the story centres around the attempt by Charlie Hard, the proprietor of the Corby House, and "Splam the Mixer," a professional race organizer, to fix the results of a foot race for the "Corby Cup" to be contested on Dominion Day. Hard is portrayed as "a born gambler, that curious cold-blooded type, without nerve, or with nerves of steel, that makes gambling a success financially." He is "poker faced, in his countenance," and "no heat of passion thawed that pale-skin coldness." His only credit in the "moral ledger" is that he took the most solicitous care of his mother.

Charlie Hard and the Mixer agree to plant a professional runner in Petroville. Hard was eager to "pull one over," on oil producer and hotel proprietor, Ted Corby, who was probably modelled after Harrison Corey, an oil producer and proprietor of the Corey House. A professional runner was obtained in Detroit, and arrangements were made for the runner to have a job in Petroville and live in the community for one month before the race. A residency requirement was stipulated in the rules for the Corby Cup to exclude professionals who might drop in for the day's sport.

A professional runner, Jimmy Clancy, was smuggled into town. Charlie Hard arranges for Corby to find out that Clancy is a ringer by having Clancy set his running shoes outside his hotel room door. At one o'clock on Dominion Day, the town brass band forms a circle in front of the Corby House and plays the Maple Leaf. Next, the band marches towards the racecourse, followed by a cavalcade of buggies and men on foot. The race begins amidst numerous allegations of fixing. In the end moral right triumphs and the ringer is defeated.

Following the publication of Fraser's story, Frank Simmons, a former resident of Petrolia, wrote to the editor of the *Advertiser-Topic*. In his letter Simmons remarked that the story had revived old memories of foot races held in Petrolia during the 1860s. He recalled a distant relative by marriage, Sam Wright, who defeated the acknowledged local champion in a middle-distance race in the summer of 1869. Wright went to Oil Springs from Plainfield, Illinois, in 1855, and relocated to Petrolia in 1867. Simmons remembered Sam Wright as his "ideal of an athlete;" "he was in no sense a ringer," but a true amateur.[65]

Local newspaper accounts attest to the popularity of foot racing among young men and boys in Petrolia during the 1860s and 1870s. They wagered on the races held on the half-mile track at the East End of town. The editor of the *Topic* was critical of the rowdiness: "The amount of yelling and swearing indulged in, together with the almost naked state of a great many of the contestants, is enough to try the patience of a saint." The editor urged them to try to find some other place to stage foot races, and "give the citizens a little peace in that neighbourhood."[66] Alongside this rough masculine culture, usually associated with working-class tavern culture and male sporting culture, the emergent middle class, most of whom were also members of fraternal orders and the literary societies, used civic celebrations, amateur sport, and assemblies to create their own culture of respectability and refined sociability.

In November 1867 a group of leading oil producers, businessmen, and professionals – B.S. VanTuyl, George Moncrieff, Romeyn Lawyer, Frank Smith, William Ewing, and Richard Herring – met in the United States Hotel and inaugurated the Petrolia Assemblies. As historian John Kasson explains, the expressive, often rowdy, male-dominated assemblies that played so conspicuous a part in town and city life in the first half of the nineteenth century were transformed into disciplined, passive, and segmented gatherings for the middle classes in which women figured prominently. The new, more exacting terms of sociability and control of emotional expression, formed in the middle-class private sphere of the home in the evolution of dining manners and parlour etiquette, transformed the public sphere as well. A new middle-class ideal was inscribed upon the community: orderly, regulated, learned, prosperous, and "civilized."[67]

Petrolia's middle classes were preoccupied with seeing and being seen in public. The Petrolia Assemblies were staged events, with men and women adopting distinctively different class identities based on a gender-specific protocol. Each year a group of "fit and proper persons" from the town's male middle classes were appointed as patrons to arrange the assemblies for the season. The patrons were prominent merchants, professionals, and oil producers, namely, Harrison Corey, A.C. Edward, Edwin Kerby, Henry W. Lancey, Frank Smith, J.H. Fairbank, Romeyn Lawyer, Robert D. Noble, J.L. Englehart, Albert

Scarsbrook, Benjamin S. VanTuyl, Stephen F. Griffiths, George Moncrieff, Peter Campbell, and Leonard B. Vaughn. The wives of the patrons were designated lady patronesses. It was the role of the patrons and lady patronesses to make sure that the "high tone and character" of the gatherings was maintained, and to assist the younger generation in learning the rules of proper etiquette and refined sociability. The assemblies, with their established codes of behaviour, were exclusionary. They were essentially fancy balls for the town's upper middle classes and they excluded the working class and the more marginal members of the middle class, although these could observe from the gallery. Assemblies, as Kasson argues, also acted as a check against a fully democratic social order by supporting the interests of the middle classes, and their institutions of privilege and structures of domination.[68]

The assemblies were a mixture of spectacle and restraint. In December 1891 two working-class women, Jessica and Belle Hazleworth, visited the Petrolia Assemblies held in Victoria Hall, as correspondents for the *Topic*. Because Petrolia had international connections based on the technical skills in oil drilling of its renowned foreign drillers, the assemblies attracted visitors from as far away as Austria and New York, as well as from Detroit, London, Chatham, and Sarnia, representing the *crème de la crème* of society. From the gallery of the opera house, occupied by a bevy of "embryo belles and their chaperons," and other observers like themselves, the women described how the event was staged. The ladies emerged from their dressing room, clothed in elaborate costumes. For women, the crucial medium of display was costume. They were attired in opulent gowns, and the two correspondents were undecided as to who was the "belle of the ball." Two potential candidates for the distinction were described as "dainty," "graceful," and "dignified." Their elegance, in part, was derived from an ability to manage a long train without tripping.[69] The Petrolia Assemblies thus provided a public sphere, into which respectable women not only could enter but had to enter, in order to participate in the culture of refined sociability.

According to Jessica and Belle Hazleworth, the men emerged from the gentlemen's dressing rooms in "a magnificent display of white linen, gloves and swallow tails or Prince Alberts." The ladies were then led by their gentlemen friends to the seats placed along the sides of the room. A gentlemen was required to spend the evening attending to the ladies in his charge; two at the most, according to the rules of etiquette. A program of dances was provided. A gentleman arranged to dance with a lady by asking to sign her program. The evening's dancing began with a Grand March and Saratoga Lancers, followed by a waltz. During the grand march the elegant and costly costumes of the ladies were displayed in all their brilliance, contrasting with the conventional black dress of the gentlemen.

Occasionally the rules of etiquette were transgressed with the justification that the custom of the new country was different. Following the premier assembly for the season in December 1893, the editor of the *Advertiser* complained that the rules of etiquette had been abandoned by those gentlemen who engage six dances or more ahead, and by gentlemen who escort their ladies to their seats and then abandon them to find other ladies for the first dances. The editor remarked: "There is only one etiquette, and if we are to maintain a society standard the strict rules of etiquette must be observed, more especially at a public assembly, and our assemblies, broadly speaking, are public assemblies."[70] The elite, therefore, had to be seen adhering to the rules of etiquette, where discipline and decorum guided sociability, thereby reinforcing middle-class domination.

The gendered rules of etiquette were turned on their head at the New Year's Assembly when it fell on a leap year. Belle Hazleworth described the assembly held on 12 January 1893, where ladies and gentlemen reversed roles and costumes. The invitations were printed on calico, and silk dresses were abandoned for calico and cotton. Calico was used in the dresses of working-class women, and thus symbolized a transgression of the social order. The gentlemen emerged from their dressing room, and were escorted to their seats on the side of the room by the ladies. The ladies, Belle writes, "were not at all awkward in assuming their duties as escorts," and went about "their own sweet wills" arranging dances. The ladies apparently were eager to engage in "the Leap Year privileges," while the men, according to the woman reporter, were decidedly bashful.[71]

Civic celebrations, especially the Queen's Birthday and Dominion Day, with their colorful parades and sporting events, were another means by which Petrolia's male middle-class businessmen used their privileged access to the public sphere to legitimate their status as community leaders. Parades, as the writings of cultural historians have demonstrated, were an important form of communication in nineteenth-century North America. Not everyone had the same access to the streets. "Images of social relations," historian Susan Davis explains, "were filtered through a complex process of inclusion, exclusion, influence and planning, until the parade expressed power and special interest more than unity and consensus."[72] During the 1880s and 1890s, civic celebrations, and the public display of order and decoration in street parades, were used by middle-class businessmen, who organized, promoted, and carried out the events, to articulate respectability and to promote a community identity built around the oil industry. While some elements of the disrespectful and the disorderly traditions of the rough culture of the working classes were incorporated into these parades, their only access to street theatre was through carnivalesque mockery of the powerful. Civic holidays also provided nineteenth-century townspeople with a rare opportunity for leisure pursuits, and sporting events were used to articulate middle-class masculine ideals of amateurism and respectability.[73]

In 1878 the Queen's Birthday was celebrated with a Callithumpian procession that paraded the streets of Petrolia banging pots and pans, and in the tradition of anti-authoritarian burlesque ridiculed the established social order. The parade was followed by a program of sports and games at the new driving park in the adjacent community of Marthaville. A baseball game between the "East-end" and the "West-end" men of Petrolia was organized. Many of the town's residents, however, took the opportunity to catch the 7:30 a.m. G.W.R. excursion to Sarnia, where they enjoyed the holiday swimming and fishing.[74] For the remainder of the nineteenth century, the Queen's Birthday celebrations in Petrolia were usually quiet events in comparison to the Dominion Day celebrations, with many citizens choosing to forgo any formal public celebrations by taking advantage of railway excursion rates to take the day trip to Sarnia.

The revelry of the burlesque street parade that marked the Queen's Birthday celebrations in Petrolia in 1878 was not replicated in Dominion Day celebrations held in the 1880s and 1890s, although Callithumpians were sometimes incorporated into the parades organized by the town's businessmen. Petrolia's Dominion Day celebrations were reported in detail in the town's newspapers. Businessmen and oil producers designed the Dominion Day spectacle to conform to their own well-established ideals of respectable white middle-class manhood. Dominion Day parades were a legitimation of the power of the town's propertied middle class and communicated the ideals of Anglo-Protestant middle-class respectability using patriotic and military ritual.

A committee of socially prominent businessmen, who were also active in the town's associational life and held public office, planned and executed the Dominion Day celebrations. In 1885 Petrolia's Dominion Day celebrations began in the morning with a parade. The procession lined up behind the marshals: oil producers Frank Smith and James Peat, book and stationery merchant E.A. Archer, and explosives manufacturer Richard I. Bradley. The parade marched down the streets where the middle class lived and did business, along Petrolia Street to First Avenue, around First Avenue, Pearl Street, and England Avenue to Petrolia Street again, finishing at the market square. According to the *Advertiser*, "the majority of the residents of the town had the privilege of seeing the procession," although the working-class families who lined the parade route did not have the same access to public space as the middle-class organizers, as they were given only limited access.[75] The affluent middle-class men who organized and paid for the event projected a dignified and patriotic image through the orderly public parade, thereby confirming their position of privilege in the social order.

Following the marshals in orderly procession were: the Oil Springs Band, members of the local militia, the fire-hose cart, the Andes Fire Company, the Andes Steamer, and the Phoenix Steamer. Petrolia's Dominion Day parades served as a powerful, recurrent context for volunteer self-presentation, and embodied the middle-class image of decorum and manly self-control that parade organizers were eager to communicate to the public on the sidelines. A trades procession completed the 1885 Dominion Day parade. This commercialized parade was a form of civic boosterism aimed at highlighting the importance of industry and commercial enterprise to the town. Leading the trades procession was a float representing the *Advertiser* office, complete with a press in operation and a case of type on the rack. Atop the Tecumseh House float was a dining room in operation complete with "colored waiters." The trades procession, as Heron and Penfold indicate, put craftsmanship on display, and parade-makers used these animated floats to highlight skilled workers' identities as citizens in a democratic country, who also conformed to the well-established conventions about orderly processions. Women, non-whites, and skilled workers were excluded from the planning of the Dominion Day parades, but while on display they were given limited access to public space. These exhibitions of workmanship bore the employers' name, and local manufacturers saw the pageants as a form of advertising. "The working-class paraders," Heron and Penfold write, "thus simultaneously demonstrated their technical skills and advertised their employers' products."[76]

The trades procession in Petrolia on Dominion Day 1885 was infused with a political message aimed at the syndicates of oil producers, which many small producers believed were detrimental to the interests of the local oil industry. Displayed on the "Syndicate Wagon" was a company of men living in luxury. The next float depicted the "Poor Producer" represented by the character "Greasy Charlie," sitting at the foot of a derrick in pitiful shape. Behind the "Poor Producer" rode producers who did not belong to the syndicate, imagined as a "plug hat gentry smoking good tobacco."[77]

Several years later, in the Dominion Day parade for 1892, a burlesque of the future of the Petrolia oil industry was presented on two floats. One float was a prophecy of what Petrolia would be like in 1895, "if we make good oil." The wagon was a strong well-made vehicle, drawn by a span of well-fed and well-groomed horses. George Palmer, a local operator rode on top of the wagon, symbolizing the prosperous oil producer, "well dressed, with a bouquet in his buttonhole and holding an umbrella to keep the sun from tanning his carefully preserved complexion." As the *Advertiser* remarked, "the entire 'get-up' bespoke the highest grade of prosperity." This was followed by the dilapidated rig suggesting that Petrolia's absolute downfall would occur if refiners made an inferior grade of oil. The driver was clad in rags. "His appearance," the *Advertiser* noted, "was that of a filthy, neglected loafer, devoid of civilization or the slightest pretense to ambition." The horse he drove was blind and lame, and the wagon was a dilapidated old dray on three wheels. The oil barrels were old tubs that apparently had done years of service as old swill barrels. The message that the middle-class businessmen and parade organizers intended to convey was clear: the future of Petrolia, and the Canadian oil industry, called for local refiners to put only the very best quality of oil on the market.[78] The identity of the town was shaped by the oil industry.

Civic ceremonies in nineteenth-century North America displayed the public in deeply gendered imagery. Parades were encoded as masculine, and women, as Mary Ryan and Cecilia Morgan have argued, were cast as the dutiful spouses of male elites or as demure symbols of civic virtues.[79] Women could decorate public ceremonies, but they could not plan them. Women symbolized the virtues of patriotism and nationalism in Petrolia's nineteenth-century Dominion Day parades.

On Dominion Day 1892 Miss Jennie McDonald appeared on a beautifully decorated float as "Miss Canada," the guardian of the continuity of the nation. A court of young ladies accompanied her, consisting of the daughters of some of the town's most prominent businessmen and oil producers.[80]

In small towns and cities throughout nineteenth-century Canada, Victoria Day and Dominion Day celebrations marked the beginning of the spring and summer sporting seasons, and sporting events were an important component of the festivities.[81] Petrolia's Dominion Day celebrations opened the summer season of horse racing at Greenwood Driving Park. On 31 January 1881 letters patent incorporating the Greenwood Driving Park Association were taken out by three Petrolia oil merchants – Frank Lewis, Jacob L. Englehart, and Richard I. Bradley – Petrolia hotel keeper Joseph Huggard, Enniskillen oilman Elijah Coreyell, and farmer William Anderson. The men petitioned to purchase twenty-five acres of land for the purpose of improving the breed of horses, holding fairs, and fostering athletic and manly sports. The creation of a local Driving Park Association was intended to elevate the moral tone of horseracing, which was often an avenue for gambling.[82] The middle-class organizers of civic celebrations took the popular athletic games of the working classes and organized them into a program of athletic events for a cash prize. The activities included novelty athletic events such as fat man's races, three-legged races, and sack races. Team sports were also part of the festivities.

The sporting culture that developed in latter nineteenth-century Canada, as sport historians have taught us, was dominated by team sports.[83] The new sport culture was largely the creation of middle-class men. Rationalized, codified sports embraced British ideals of manliness and gentlemanly conduct, including the "character-building" qualities of amateur team sports. Over the course of the nineteenth century, sport became enmeshed in the project of building local community identities and defining Canadian national identity within the British Empire. Sport emerged as an institution of social fatherhood by legitimating the status of middle-class men as community leaders. As Varda Burstyn argues, the largely successful exclusion of women from sport in the late nineteenth century reinforced the idea of separate gender spheres and masculine entitlement to public power and the domestic service of women.[84]

During the 1880s and 1890s inter-town baseball matches were usually incorporated into Petrolia's Victoria Day and Dominion Day festivities, with the Petrolia Baseball Club taking on a team from Sarnia, Watford, or London. As in other Canadian towns of the period, baseball became the game of choice in Petrolia over cricket and lacrosse, because of its character-building values. Baseball was attractive to middle-class reformers. It brought into play the manly virtues of courage, strength, agility, teamwork, decision-making, and foresight. Baseball, as Colin Howell and Bryan Palmer point out, was also incorporated into nineteenth-century working-class culture. Baseball required no elaborate playing-field, little equipment, and only a limited amount of instruction before a working man could enjoy the game. Frequent lulls between pitches also allowed for socializing among players and spectators, while the more physically intensive games of hockey and rugby did not.[85]

Although Petrolia had both a cricket club and a lacrosse club in the 1880s, civic boosters viewed inter-town baseball matches between representative teams as a way of shaping the worth of the community. On Dominion Day 1877 the Petrolia baseball team travelled to Watford to play the local team. The Petrolia's defeated the Watford Blue Stockings by a score of 9 to 7. Prior to the match an agreement was made between the two clubs that five innings would conclude the game. When the Watford team had not "scooped" the Petrolias after five innings they called the game a draw even though the umpire had given his decision in favour of Petrolia. The Petrolia *Advertiser* criticized the Watford team for its lack of manliness on the playing field. "In fact, they were beaten," the *Advertiser* stated, "and had not the courage to take their defeat with that good grace which characterizes all players who have a wish to hold the respect of those interested in the game of baseball." The worth of the community was tied to the masculine respectability of its players on the baseball field. "Our boys say they will, in case the 'Blues' come here to play, at least extend to them a show of decency and welcome, which they assert was not accorded to them on the occasion mentioned."[86]

In Ontario the bicycle craze began in the 1880s. The first bicycles, penny-farthings with their large front wheel, were expensive and beyond the means of any but Petrolia's wealthiest citizens. On 13 July 1888 the Petrolia Wheel Club held its first meeting, and Dr John Dunfield was elected president. With the invention of pneumatic tires and the inner tube by John Dunlop in 1889, and the advent of the mass-produced "safety-bicycle," the bicycle craze hit Petrolia. A "Bicyclists' fancy drill" and distance races were part of the Dominion Day program in 1890, with the *Topic* reporting that several "headers" were taken by wheelmen.[87]

Another club, The Imperial Bicycle Club, was organized in 1892 by several of the town's leading businessmen. George Moncrieff was chosen honorary president and oil producer J.L. Englehart, honorary vice-president. A.E. Fitzgerald offered a gold medal to the member who accumulated the most miles on club runs during the season. Bicycle club excursions, with their precision military-style lineups under the direction of annually elected officers, complete with military designations and a "whipper-in," emerged as another popular type of parade and another form of civic boosterism in Petrolia. In June 1893 the *Advertiser* reported that the Imperials covered 70 miles on their Sunday run to Marine City, Michigan, despite roads that "in some instances were well nigh impassable" after the rains of the two previous days.[88]

Local interest in bicycling intensified when Harrison Corey's son Bloss, a member of the Imperial club, began attending races in Sarnia and won the county championship in 1891. Amateur bicycle races, conducted under the auspices of the Canadian Wheelmen's Association, were thought to inculcate the virtues of respectable manhood. As Bloss Corey's victories in amateur races continued he was described with pride in the Petrolia *Advertiser* as "Our Coming Man."[89] The races were exclusively male events, although the presence of women as spectators confirmed the status of middle-class respectability. In 1894 the executive of the Imperial Bicycle Club successfully lobbied the town council for a meet to be held on the August civic holiday and for the construction of a new bicycle track. According to the *Advertiser*, the event would serve to "advertise the town and be an inducement for wealthy industrial firms to locate their works where a spirit of enterprise is apparent."[90]

By the early 1890s the parameters of middle-class propriety had shifted and ladies began to ride bicycles. Each week the *Advertiser* reported the names of new cyclists, including the names of women and the make of bicycle they were riding: "Mrs. R. Herring rides a handsome No. 2 'Fleet';" "Mrs. Ernie Fitzgerald has exchanged her Quadrant for one of Hyslop & Co's No. 2 'Specials';" "Miss Emma Bickle and Miss Tina Polley are riding a Monarch." Local debates over appropriate women's clothing for cycling suggest that cycling provided a public space in which notions of femininity and appropriate appearance were contested. In the summer of 1895 the *Advertiser* suggested to the public that bloomers were "the only proper and least suggestive style of apparel for lady cyclers to wear," and a sum of $50 was pledged to the lady who donned bloomers first. On August 1st the *Advertiser* reported that "the honor of the emancipation of the sex belongs to Miss Emma Bickle."[91]

The public sphere in latter nineteenth-century Petrolia was the terrain of middle-class men, although the fundraising work of women was vital to its survival, and well-articulated notions of middle-class femininity helped to shape the middle-class public sphere of refined sociability.

CHAPTER FIVE

"Some Adventures of the Boys": Enniskillen Township's "Foreign Drillers," Imperialism, and Colonial Discourse

In December 1873 the first crew of Enniskillen Township's foreign drillers left Petrolia. The drilling crew consisted of oil engineer Malcolm "Mall" Scott, scaffold man William Covert, and drillers Joshua Porter and Edward Cook. Their destination was the Dutch colony of Tjibodas, at the foot of the volcano Tjarema, south of Cheribon in West Java. In a scene that would be repeated numerous times over the next fifty years as drillers from Enniskillen Township departed for what were referred to locally as foreign fields, friends and family gathered at the train station to see the men off. The men departed to the strains of the old Scottish song "Will Ye No Come Back Again" played by the town band.[1] Ten months later the *Petrolia Advertiser* published extracts from a personal letter from one of the drillers, Joshua Porter, to his brother Joseph. The letter was composed on 17 July 1874, shortly after the crew of drillers arrived at Tjibodas. Porter's verbal painting of the landscape produced for the home audience a sweeping visual mastery of the scene, which Mary Pratt describes, as "the monarch-of-all-I-survey scene."[2] He depicted East India as showing much promise for Western development, "being covered with rich vegetation, from the mountain top to the waters edge." Consistent with the narratives of nineteenth-century European travel writers, Porter's mapping of Batavia, Java (the present day city of Jakarta, Indonesia) reinforced European racial hierarchies. "The hotels are numerous and grand buildings, always back from the streets and surrounded by gorgeous grounds, giving an appearance of rich private residences," Porter reported. "The European stores are built in the same style, but the Chinese shops and native markets are a curiosity."[3] The mapping of non-white Eastern others as a "curiosity" highlights the "mastery" of white Western men and disavows the agency of non-white Eastern Others. Porter highlights the triumph of Western civilization over the wilderness. "Our next travelling was done in carriages, passing many plantations, sugar factorys [sic], large tracts of rice fields, and peanuts, (where a large business is done in manufacturing oil from peanuts) we, also, saw extensive patches of sweet potatoes."[4]

Paradoxically, while Porter constructs racial hierarchies as a means of establishing cultural difference between the white drillers and the native Others, his narrative is one of appropriation where he dominates by gathering the colonized into "civilization." In the latter part of his letter, Porter describes their quarters in Tybodus. "Our bamboo houses are comfortable, and servants – two men and a woman – at 10 cts a day each." There are, however, underlying tensions evident in Porter's rhetoric of appropriation and domestication. "We could fare very well, if our meals were cooked right, that is, such as in our own country," he complained.[5]

In a 1924 *MacLean's* article commemorating the golden jubilee of foreign drilling, Victor Lauriston, a journalist and local Lambton County amateur historian, dubbed Petrolia "The Town of World Travelers." According to Lauriston, 134 drillers from Enniskillen Township went on exploration and drilling expeditions to foreign fields between 1873 and the outbreak of World War I.[6] Many of them went abroad more than once. A little known aspect of Canadian history is that the foreign drillers from Lambton County, many of them from Petrolia and Oil Springs, provided the skilled labour and technical expertise necessary for the development of the global oil industry. The vagaries of the local oil market and the depletion of the oil resources in Enniskillen Township by the late 1890s forced many skilled oil drillers to seek employment in foreign fields. In an era when only the wealthy, soldiers, and diplomats travelled abroad extensively, Enniskillen drillers travelled to the oil-rich fields of Java, Borneo, Sumatra, Persia, Galicia, Germany, India, Newfoundland, Russia, Italy, the West Indies, and the United States. Virtually anywhere in the world where oil was discovered, Enniskillen's drillers provided the skilled labour and technical expertise.

In the latter half of the nineteenth century petroleum was used primarily for illuminating oil. It was also used in the manufacture of benzene, gasoline, kerosene, paraffin wax, asphalt, and lubricating oil. Petroleum lubricants greased the machinery that powered the industrial revolution. The demand for crude oil increased significantly in the early twentieth century, particularly with the introduction of the automobile as a means of mass transportation. The exploitation of oil resources fuelled European imperialism. At the turn of the century virtually all of Britain's oil came from the U.S. and Russia, with smaller amounts coming from the Dutch East Indies and Romania. All of these countries were outside Britain's control. At the outbreak of World War I in 1914 the world's largest oil concerns were American, Russian, and Dutch. One of the British Admiralty's primary concerns was to secure an adequate supply of fuel for the navy, and the British government framed an oil policy to win control over the oil supplies in the Middle East.[7] Beginning in the early 1870s drillers from Enniskillen Township journeyed to these remote areas, bringing their technical knowledge of "the Canadian drilling system," along with the components of the drilling rigs, which were often manufactured locally at the Oil Well Supply Company.

Enniskillen foreign drillers, as both British subjects of the settler colony of Canada and agents of imperialism, went about constructing a coherent representation out of the strange realities they confronted while working in the colonial oil fields. The foreign drillers provided the skilled labour and technical expertise necessary for the development of the colonial oil fields. They were also bearers of class, gender, and racial privilege in the oil fields where they were sent to work, as they went about reinforcing the project of European capitalist imperialism while simultaneously disavowing the agency of non-white native "Others." For a significant proportion of Enniskillen Township's foreign drillers, working for a large international oil company provided them with the opportunity to secure supervisory and management-level jobs and a position of class privilege while living abroad. This no doubt would have eluded them had they spent their entire lives working at home in the local oil industry. Enniskillen's foreign drillers became part of an imperial "overclass" by virtue of their "whiteness," "Britishness," and technical expertise in the mining and refining of petroleum.

Recent rethinking of imperial history and efforts to integrate imperial history with British domestic history have benefitted from interdisciplinary exchanges, particularly the insights of post-colonial scholars. Until recently, Antoinette Burton remarks, "the rule of thumb in British history has been to map a set of quite differently imagined communities: 'home' on the one hand and 'empire' on the other." Home and empire have traditionally been constituted as separate and distinct spheres, with home being the source of Britishness, progress, and civilization, and Empire as the other side of the world, the 'dark continent,' the as-yet undomesticated space of cultural backwardness.[8] The so-called "new" British historians have instead suggested that the relationship between home and empire was dialectic rather than dichotomous. Historians such as Antoinette Burton, John Mackenzie, Laura Tabili, Anne McClintock, and Catherine Hall have illustrated that empire was an integral part of British social, political, and cultural history. "Domestic" British culture was thoroughly influenced by imperialist ideas and, with it, the meaning of racial difference.[9] In Canadian historical writing, Carl Berger has argued that in English Canada, between Confederation and the outbreak of World War I, it was possible to be both a nationalist and an imperialist.[10] In his presidential address to the Canadian Historical Association in 1993, Phillip Buckner stated: "Canadian historians locked themselves into a teleological framework which is obsessed with the evolution of Canadian autonomy and the construction of a Canadian national identity and thus downplayed the significance of the imperial experience in shaping the identity of nineteenth-century British Canadians."[11] Through travel writing and work abroad, a dialectic between "home" and "away" and connections between "nation" and "Empire" were forged by Enniskillen's foreign drillers. The colonial oil fields also became a space for the re-invention of Victorian ideals of domesticity. Thus an argument can be made for the "peculiarities of the Canadians" both as British subjects of a white settler colony and as colonizers in the colonies of oil exploitation.

Post-colonial theorists have argued for a re-examination of colonialism, beyond a simple analysis of machines of imperial power exercised from European metropoles, to incorporate the interactive, social dimensions of colonization that took place in the colonies of exploitation.[12] In her study of European travel writing, Mary Pratt foregrounds the interactive dimensions of colonial encounters with an emphasis on how subjects are constituted in and by their relations to one another using the idea of the "contact zone." The "contact zone," Pratt writes, refers "to the space of colonial encounters, the space in which peoples geographically and historically separated come into contact with each other and establish ongoing relations, usually involving conditions of coercion, and radical inequality, and intractable conflict."[13] Thus, the "contact zone" provides a useful way for analysing how Enniskillen's foreign drillers went about constructing the domestic subject of European imperialism as colonizers in the colonies of oil resource exploitation.

Some of the travel letters written by Enniskillen's foreign drillers were published in the two local newspapers, the *Petrolia Advertiser* and the *Petrolea Topic*, later merged as the *Advertiser-Topic*, during the heyday of foreign drilling between 1874 and 1930. The letters were sometimes addressed to friends or family members, but more often were in the form of letters to the editors of the two weekly newspapers. In November 1897 Richard Herring, editor of the *Advertiser*, issued a call for letters from subscribers living in foreign countries. "We would take it as a favor," Herring wrote, "if they will write to this paper, briefly stating their impressions of their present places of residence, and also give our readers an idea of how those places compare with their old town, Petrolia ... We are confident that the letters we will receive will be well worth reading and will verify the statement that 'nobody who ever lived in Petrolia failed to return.'"[14] The travel letter was perhaps the simplest form of travel chronicle and appeared widely in nineteenth-century North American newspapers.

Travel writing gave Enniskillen's foreign drillers the opportunity to produce themselves as authorities on travel to distant lands and other topics of general interest to their homebound neighbours in small-town rural Ontario. The travel chronicle, as William Stowe explains, testifies to a privileged experience and constitutes an authoritative voice.[15] Enniskillen's foreign drillers were empowered by the experience of travel and travel writing, and had a coveted role in the local community. They had high-paying jobs in comparison to men who remained at home, and they had the opportunity to travel abroad, which allowed them to claim respected cultural roles in the community. The letters composed by Enniskillen drillers working abroad can be distinguished from fiction by the expectation of their grounding in historical actuality. Despite this quite conventional expectation, the use of symbols, myths, metaphors, and other rhetorical strategies more often associated with fiction than travel writing can be used to establish the connections between home, nation, and empire and to establish hierarchies of race, class, and gender power. The letters are a kind of conversation – albeit one-sided – where the relationship between colonizer and colonized was one of "co-presence, interaction, interlocking understandings and practices."[16] Quite possibly the drillers altered their narratives to conform to the expectations of their readers at home, although this could not be determined for certain. Used alongside extant memoirs, diaries, and personal family letters, the drillers' travel letters are useful for identifying and mapping the basic tropes of colonial discourses.

ENNISKILLEN TOWNSHIP'S FOREIGN DRILLERS

The movement of Canadian drillers to foreign fields received its greatest impetus in the 1880s from the enterprise of Petrolian William Henry McGarvey. McGarvey was born in Huntingdon, Quebec, on 27 November 1843. In 1857 the McGarvey family moved to Wyoming, approximately seven miles to the north of Petrolia. In 1861 McGarvey relocated to Petrolia, where he opened a general store known as The Mammoth Store. Following the incorporation of Petrolia as a village on 13 December 1866, McGarvey, who was only twenty-three years of age at the time, became the first reeve. He held the position until 5 March 1867. In 1876 he was elected mayor and in 1879 he became warden of the County of Lambton. A staunch Conservative in politics, McGarvey headed a federal royal commission to survey the mineral resources of the north-west in 1880. The following year he was appointed manager of the Continental Oil Company, where he became acquainted with J.S. Berheigm, an English oil operator. In partnership with Berheigm, McGarvey drilled for oil in the German province of Hanover. This venture was unsuccessful, and in 1882 McGarvey and Berheigm proceeded to Galicia and Romania. McGarvey introduced the Canadian system of drilling in the oil fields in Galicia.[17]

The Canadian drilling system, also known as pole tool drilling, was perfected in the oil fields of Enniskillen Township during the oil boom days in Petrolia in the late 1860s. The surface or flowing wells were spent within a few years after oil speculators began extracting the oil resources in the region. A new science and trade in oil drilling emerged to meet the need to drill deeper into the rock for oil. The method of sinking the early oil wells in Enniskillen Township in the 1850s and early 1860s was to dig and crib a four-foot shaft down to rock, then sink the well through the rock by driving the drill into the ground. These first wells were "kicked down" by manpower using spring poles. This method was soon replaced by steam-powered drilling rigs, known as the Canadian rig. A bit or cutting tool in the form of a chisel screwed to an auger stem was connected to "jars" or slips, which permitted a delayed power motion on both the upward and downward strokes. Unlike the American, or standard, system of drilling where the tools were suspended from a wire cable, in the Canadian drilling system, a series of wood tubes or wrought-iron casings were screwed together end-to-end to carry the string of tools. The rig ran the casing into the ground using a series of belts and pulleys powered by a portable steam engine driving a band wheel. The advantage of using poles was that they could hold drilling tools together better than a rope or cable while boring through rock. At first black ash from the forests of Enniskillen Township was used to construct the rods. This drilling technique was ideally suited to the wet clay soils in Enniskillen Township, where it was not necessary to bore to deep levels to find oil and where iron and steel were scarce and costly.[18]

The partnership of McGarvey and Berheigm brought in the first successful well at Waglowka in Western Galicia, then part of the Austro-Hungarian Empire, sometime in the early 1880s. By 1887 they were in a position to begin the construction of a refinery at Miariampole. In 1895 McGarvey and Berheigm reorganized their company as a limited trust company named Galizische-Karpathen Petroleum Aktien-Gesellschaft. McGarvey and Berheigm also built a drilling tool factory. At the beginning of the twentieth century the businesses at Miariampole, which at that point were under the direction of McGarvey's eldest son Fred, employed more than 1,200 workmen.[19] The oil fields in Galicia were among the largest in the world. McGarvey became a multimillionaire before the outbreak of World War I forced the closure of his wells. Before the war, McGarvey and his wife entertained members of the European aristocracy in their apartments in Vienna. On 12 November 1895 his daughter, Mary Helena, married Count Von Zeppelin, a nephew of the inventor of the airship of the same name. An account of the lavish wedding was carried in the *Petrolia Advertiser*.[20]

By the latter part of the nineteenth century Enniskillen Township's skilled oil drillers were in high demand in foreign fields. The *Topic* reported in September 1889 that while English capital had been advantageously invested in Galicia, the drilling there had been done largely by Canadians.[21] By the end of the 1880s crews of Enniskillen drillers were at work not only in Galicia but also in Romania, Italy, Russia, India, Burma, Sumatra, Borneo, Mexico, and Australia. In 1888 John Henry Fairbank testified before the Royal Commission on the Mineral Resources in Ontario that "[t]he cause of the demand is, that they have superior tools and possess superior intelligence ... The men are largely the sons of settlers and people who drifted into the oil business. The work is very hard, and requires a strong frame and a clear head. Our men become great experts at it. When they go abroad they get good wages, and expenses paid."[22]

Enniskillen's foreign drillers learned their trade locally. For instance, W.O. "Bill" Gillespie was born in Petrolia on 15 February 1878. His father looked after the oil properties of a wealthy U.S. oil speculator, a lawyer by the name of Griffiths. In 1886 Gillespie's father had an accident. When he was finally able to resume working, young Bill would help him around the wells: "get his wood for the day, put it into the engine house in front of the boiler for him so he could get along till 4 o'clock." After he was dismissed from school for the day, Bill would take over from his father until he shut down the rig at six o'clock. Griffiths eventually went bankrupt and left Petrolia. Gillespie's father then went into a short-lived contract drilling business with another local driller, Charlie Slack, before starting a cartage and surface boring business. The business was apparently successful. Tragically, the senior Gillespie was killed on 17 July 1893 while boring clay on a drilling rig in the east end of Petrolia. Bill Gillespie, who was only fifteen years of age at the time, took over his father's business temporarily. In June 1894 Bill went to work on Peter Slack's drilling rig. Charlie Keck was the driller on the rig, and Neil McGill was the engine man. Gillespie spent more than three years learning how to drill from Keck. He was paid a wage of ten dollars per well. Gillespie's career as a foreign driller began in 1903, when he was hired to go to Cuba to put down water wells. He made another trip to Cuba the following year, and then spent three years boring for the International Boring Company in Queensland, Australia. Before retiring from foreign drilling in 1926, Bill Gillespie drilled in central Africa, Upper Burma, Sumatra, and Borneo.[23]

Victor Lauriston noted that the manly characteristics of "hazard" and "muscle sense" acquired by a Canadian driller born and bred around the drilling rigs of Enniskillen Township made him a superior driller, in contrast to the Englishman who was unwilling "to take his coat off." "His work compels him to be a grimy mechanic," Lauriston wrote. "Wash that grimy mechanic, put him in evening dress, and you have – in nine cases out of ten – as fine a gentleman as Canada, or England for that matter, could wish. He has to be a thorough-going man to qualify as a driller. If by any chance a moral or physical weakling crept into the ranks, the trade would inevitably make a man of him, either that, or it would kill him."[24] In his article, Lauriston associates the masculine character traits imbued in drillers trained in the oil fields of Enniskillen Township with the burgeoning English-Canadian nationalist sentiment during the post-World War I years, and the ongoing questioning of Canada's colonial relationship with Britain.

The capriciousness of the Canadian oil industry meant that when the markets slumped local drillers were often unemployed. The first crew of foreign drillers who left Petrolia in 1873 did so in the midst of the depression that plagued the country for much of the decade, and into the next. During this period attempts were made by both local oil producers and refiners to agree on prices. Cartels were formed, which nearly always collapsed. In 1898 John D. Rockefeller's Standard Oil Company gained control of Imperial Oil and transferred the refinery to the port community of Sarnia on the St Clair River. This significant blow to the Petrolia economy was intensified a decade later when the Enniskillen oil fields began to dry up.[25] Some drillers found work in other local industries, but the lure of adventure for young men raised in rural Ontario, together with their highly specialized skills in oil drilling and the prospects of high wages, led many drillers to foreign fields. The local newspapers were filled with announcements of the arrivals and departures of drillers and their families. By-lines such as "Left For Borneo," "Leaving for India," "Among the Drillers," "Petrolia 'Jerking' Austria," and "Home from Venezuela," appeared weekly and kept the community informed of the comings and goings of local drillers. According to Lauriston, the "lure of the bit and sinker" was too strong for Enniskillen's drillers, and they would no sooner return home than they would sign another contract to drill abroad.[26]

THE TRAVEL WRITINGS OF ENNISKILLEN TOWNSHIP'S FOREIGN DRILLERS

A lengthy narrative of their journey to the colonial oil fields was included in the travel writings dispatched home by several of the foreign drillers. Their letters highlighted the boundaries between civilized, technologically advanced, Western culture and the primitive, savage wilderness of the Eastern colonies. For the drillers who went to India and the East Indies, the journey by steamship took almost two months. The drillers had to cross the Atlantic Ocean, proceed through the Mediterranean Sea, pass through the Suez Canal and the Red Sea, and cross the Indian Ocean. They usually spent a few days in

London where they finalized the terms of employment with the British agents of oil companies and had a complete physical examination.[27] The drillers used the visit as an opportunity for some sightseeing. The tourist gaze of the Enniskillen drillers was constructed around the symbols and signs of nineteenth-century British imperialism. They visited Trafalgar Square, Westminster Abbey, and the Houses of Parliament. They set their watches to Big Ben, and toured the Tower of London where they saw the Crown jewels, among them those of Queen Victoria. Madam Jarley's Wax Works and the British Arsenal were favourite attractions with the drillers.[28] In his letter to the editor of the *Advertiser*, James Rowe reported that the wax works were "very attractive, for there we find represented in wax the royal family and other persons of honor and high standing throughout the world." British imperialist discourse was incorporated into the narrative. Rowe reported that "The most pathetic figures were the Queen Regent of Spain and her son, near them was President McKinley, and he seemed embarrassed in the presence of the beautiful queen." At the British Arensal, where "one must prove himself a British subject before gaining admission," Rowe saw "all the latest improved implements of war, as well as every conceivable old kind." In London they were reunited with two more "Petrolia boys," and gave them a hearty welcome. With an expression of nationalist imperialist sentiment Rowe concluded his letter: "When journeying in these old lands we cannot forget the new land, and often feel like singing 'The Maple Leaf Forever.'"

The long journey eastward, most of it over open seas, was usually presented in the drillers' travel writings as leisurely and uneventful. Port Said, at the entrance of the Suez Canal, was the first exposure to non-European culture for Enniskillen drillers who ventured to the Orient. In the fall of 1879 R.A. Townsend was contracted by the British-owned Barangah Oil Refining Company to superintend the drilling of wells on the Barangah Islands, approximately 400 miles from Calcutta. Townsend was raised in Brant County. He went into the oil business with Melville Parker of Peel County during the oil boom in Enniskillen Township in the late 1860s.[29] Two other experienced drillers who also learned the trade locally in the Enniskillen oil fields, J.C. Crosbie and William C. Bell, accompanied Townsend on the journey to the Far East. On 8 January 1880 Townsend wrote a lengthy letter to the editor of the *Advertiser*, which was published in the newspaper in two parts.[30] Writing from the vantage point of his hotel room, he surveyed the landscape "fanned by the pure breezes from over the coffee gardens of Mocha," and remarked, "I ought to feel happy if I am not. This is fact and fancy wedded, at least this is the realization of years of dreaming, and unlike the average dream, the fact excels fancy."[31] This vision of the Orient as exotic was, as Edward Said argued, integral to the Western style of "dominating, restructuring, and having authority over the Orient."[32]

Post-colonial literary scholars have suggested that the colonizers' vision of the Orient also depended on a fantasy of seduction and eroticization in which the colonies are represented as "the harems of the West." Townsend writes that as he prowled the deck of the ship, about to experience "the dim uncertainties" of Eastern life for the first time, he happened upon what he believed to be two Eastern women "with turbaned heads and loose flowing robes." He describes the encounter as he approaches them to make his "salaams":

> With stately bearing, I approach, and with dignity and grace I touch my head, my breast and my lips – signifying that with my head I respect, with my heart I love, and with my lips I praise you. I saw at once I had made an impression, and I confidently awaited the result. It came. One of these beauties in arising, exposed about six feet of the best combination menagerie it has ever been my privilege to witness, consisting of snake skins, as malice, flees [sic], rags, gray backs, jaws, and the spicy odor peculiar to these people and this land, and addressed me in Arabic beginning with the word "Allah" and ending with "backsheesh," all of which being interpreted meaneth, Thou son of a Christian dog, your diabolical intentions are anticipated, know then that we are the sons of the Great Prophet, and are the police of yonder city sent on board to look after your health, and test your viscera by contact with opthalmia [sic], small pox, cholera, leprousy [sic], &c., and a rupee only will appease us.

The intimation of cross-dressing and disguise leads to a questioning of the masculinity of the Eastern men. "I paid the rupee," Townsend continued, "but was surprised by all this from his rank and smell."[33] The Orient in the driller's narrative vacillates between eroticization of the East and the symbolic construction of the Orient in terms of sexual danger. Townsend both delights in and fears Eastern culture. For nineteenth-century Western men, the Orient was something that had to be contained and controlled.

This erotically charged language marks the entrance of the colonizer. In this rhetorical strategy, as cultural critic David Spurr suggests, differences in power are reformulated as gender difference, and an entire people is allegorized by the figure of the female body.[34] The image of woman's sexual degeneration was used to invent hierarchical distinctions among the races and confer power among white men. In October 1897 Edward Winnett of Oil Springs left for Sumatra to assist in the construction and operation of an oil refinery. Edward and his wife Annie moved to Oil Springs from London, Ontario, in 1883. From 1885 until he left for Sumatra on 7 October 1897, Winnett operated a small boiler shop owned by his father in the village.[35] Winnett, who spent one year in Sumatra, recorded his perceptions of the experience in a diary. On November 22 the steamer *Prince Hendrick* docked in Port Said. "The first thing that caught my eye was boat loads of Egyptians, hundreds of them." Winnett wrote, "they are the worst looking lot of people I ever saw all bear [sic] naked with the exception of a filthy cloth tied around their loins and another tied on for a hat." After touring the city he concluded that, "there seems to be no civilisation there you can see ... In passing houses there you can see young girls winking and beckoning for one to come in."[36]

In their travel writings, Enniskillen's foreign drillers represented Port Said as a site of sexual danger, filth, and debasement, which empowered white Western men as agents of civilization. They were obsessed with the bodies of Arab men and women and invested them with moral value. R.A. Townsend described the Arab peoples as "filthy," "ragged," and the most "accursed lot of the 'genus homo'." "Before embarking," Townsend wrote, "we glanced at the motley crowd on the quay, men, old, gray and wrinkled women, squatting tailor-like upon the sands, with each a basket balanced upon her head, and from beneath their eyes, covering nose and mouth, a yard or more of netted filth, the famous veil of the East." Erotic fantasies surrounding the Arab women he encounters are interwoven with the trope of debasement and filth. In concluding his narrative about his visit to Port Said, Townsend comments that it is "without a regret that [he] bid adieu to this Sodom of the 19th century, this gateway of the east."[37] Subsequent drillers from Enniskillen Township gazed upon the city in a similar way. Another correspondent, Old Benwell, who provided the editor of the *Advertiser* with a narrative of his journey to Sumatra in the summer of 1898, proclaimed Port Said "the wickedest city on earth."[38]

As the ship carrying Townsend and the other two Enniskillen drillers made its way through the Suez Canal they passed a British troop ship. They exchange salutes "with the greatest flag." "I now think,"

Townsend reflected, "I never before heard a cheer and I never before was so proud and thankful for the British blood I carry about with me, thin of course, but British still."[39] Townsend's laudatory remarks about the British Empire were intended to justify imperialism. His statements suggest a more complex relationship than a simple empire/colony connection. As a colonial subject from one of Britain's white settler colonies, Townsend's British blood may have been "thin." Although he was actually a "colonial" himself, his white skin and connection with Britain as a British immigrant to Canada elevated him to a status above that of non-white natives in the colonies of oil exploitation. This suggests that the discourses of late nineteenth- and early twentieth-century colonialism are more complex than recent imperial scholarship or the writings of post-colonial theorists have implied. In the case of Enniskillen's foreign drillers, the boundaries between colonizer and colonized were never simple or dichotomous. They were subjects of one of Britain's white settler colonies while at the same time they bolstered British imperialism in the colonies of oil exploitation where they disavowed the agency of native peoples.

Not only did the drillers' narratives conform to dominant Western constructions of sexuality, they also fixed their tourist gaze on landmarks associated with Bible stories, thus furthering the project of colonizing the landscape. "We were then in the Red Sea," Old Benwell wrote in the fall of 1898 following his arrival in Sumatra. "We saw the rock that Moses smote with his staff and brought forth water for the children of Israel, also where he led them through the Red Sea, and Mount Sinai, where the ten commandments were read to the children of Israel."[40] Oil Springs boiler maker Edward Winnett was the embodiment of what historian Lynne Marks describes as the "Christian gentleman" of late nineteenth-century small-town Protestant Ontario: a man for whom Christian religious beliefs and church involvement were closely linked to a commitment to family life. Winnett attended church services held on board ship as they made their way to Sumatra. As they crossed the Red Sea he wrote in his diary: "I can see where the Israelites crossed the Red Sea. It is something most wonderful to stand on the steamer and just think of such a large body of water rolling back and allowing the Israelites to pass over."[41]

Christianity and science clashed in R.A. Townsend's narrative. As they passed the location where the Israelites supposedly crossed the Red Sea, Townsend, who was given privileged access to the Captain's charts, noted that "there is a channel mid-sea a mile or two wide and 6 to 8 fathoms deep." Townsend asked "if the children crossed here, did they have ladders with them, or were they expert at climbing?" Although Townsend doubted that any man could part the Red Sea, or that the "children of Israel" could scale its banks, his experiences as a traveller led him to retain his faith in a Christian God. "I needed not to leave the Laurentian rocks, nor the beautiful valley of the chains of lakes, nor heavens [sic] vaulted dome which covers the green fields, and broad acres of my beloved Canada, to teach me here, that their [sic] was and is a God," he wrote, "but since I have crossed many seas, since a new hemisphere, a new earth, has been passing beneath my gaze, like a panoramic ribbon, I have added lessons, and mine eyes have seen His glory in other forms."[42]

As the drillers approached their destinations their travel writings tended to take the form of boys' adventure stories, where the young heroes were fine exponents of British manliness and muscular Christianity. They must confront the unknown "wilderness" of the jungles of the Orient or face pirates on the Arabian Sea. This literary genre was no doubt familiar to a generation of young men raised on a steady diet of adventure stories published in *The Boy's Own Paper*, or versed in the poetry of Rudyard Kipling and Henry Newbolt.[43] Challenge and response were part and parcel of the project of Empire-building.

The language of "wilderness" and the perils of the unknown in the drillers' narratives takes on an originating authority and confers upon them power to represent themselves. It is up to the drillers to systematize and bring order to the wilderness. The drillers' adventure narratives further conferred upon them the status of conquering heroes among their friends and neighbours at home. On 1 March 1899 the *Advertiser* published a letter from A.B.C. under the heading "Some Adventures of the Boys." In his letter the correspondent describes the dangers encountered as the drillers made their way through the jungles of Sumatra. "When we left the launch it was a queer experience to be sailing up a small river in a small boat in the wilds of Sumatra, with nothing but woods on either side of you, and it so thick that the eye could not penetrate more than two or three yards, and in some places not so far ... The stream is very crooked, and a boat ten yards ahead of you could be easily lost track of, but returning suddenly to view, only to be lost again in another turn either to the right or left." A.B.C. pronounces the pattern of the growth of the trees alongside the river as "strange," in comparison to those at home. "There is no grass nor anything else but trees grow on the banks of the stream," he observes. "It is a strange sight to see, and what we would call the trunk or stump of the tree stands several feet from the ground, and the roots branch out on every side – big ugly claws."[44]

At the same time, the driller reinscribes the imperial social order with its hierarchy of racial privilege. A.B.C. reveals that "Cooleys" transported their luggage and supplies up the river. Following their arrival at the drilling site, the crew of Canadian drillers watches the natives work for a time. According to A.B.C., "they did not do very well because they did not understand the drilling business at that time," and they required the knowledge and expertise of the "civilized" Canadian drillers to develop the petroleum resources.

Confrontations with wildlife were integrated into the drillers' adventure stories. John Collins wrote about his encounter with a tiger roaming around the drillers' quarters in Kotei, East Borneo, in May 1900:

> I was awakened at 3:00 am this morning on hearing the dogs barking and although I tried to prevent them, I found to my surprise that a tiger was underneath the mess room. I immediately returned to my room and got my gun, and found the animal still crouching in the same position on my return. All I could distinguish were his eyes, as the morning was very dark and cloudy. So I fired in that direction, and the shot had the desired effect, as the tiger rolled over and died immediately. It measured six feet from tip to tip, and was three feet high, and was, I believe, a young animal in prime condition.

Collins further indicated that he hoped to have the skin cured, so that he might show his "trophy" to his friends when he returned to Petrolia.[45] Other drillers provided readers back home with stories of their heroic encounters with poisonous snakes, some reportedly twenty-six feet long!

For the foreign drillers who journeyed to the East Indies and British Burma in the late nineteenth-century the curios mailed home to Petrolia added to the "authenticity" of the drillers' narratives of adventure and contributed to the invention of imperial masculinity. In May 1880 R.A. Townsend sent the skin of a "very" poisonous snake to Richard Herring, editor of the *Advertiser*. Accompanying the snakeskin was a note informing Herring that the snake was of the "Tie Prolongi" species, and that this species found "it rare sport to chase a Christian at sight." Some of Townsend's men had apparently killed the snake on the roof of the cookhouse. Townsend concluded his letter with a postscript informing Herring that he need not be afraid to handle the snakeskin, as it was not poisonous.[46] Occasionally the two local newspapers published lists of "curiosities" sent home to local citizens. Thus, the colonial adventure was consumed imaginatively when the drillers wrote their own scripts and literally when the drillers sent curios to friends and family back home.

Adventure narratives composed by Enniskillen's foreign drillers continued to appear in the local newspapers into the 1920s, although by this time travel abroad was becoming a regular occurrence for local drillers and their families. In May 1922 Oil Springs driller W.C. Parker wrote that they were attacked by pirates in the Arabian Sea as they made their way to Persia to drill for the Anglo-Persian Oil Company. The pirates, described by Parker as "dusky savages," appeared on the deck of the steamer "armed with axes, piles and swords or cutlasses similar to those used in the days of Drake and Nelson." Twenty-year-old Ernest Kells, who started working on the drilling rigs in Petrolia at the age of seventeen, is presented in the narrative as the masculine hero. "Kells being a cool headed, resourceful young man," Parker wrote, "calmly stooped and picked up a bar of 1 inch iron which lay handy, with which he attacked three of the pirates." Kells hit one of the three pirates, and the other two quickly retreated. By this time the crew of the steamer, forty-five men in all, had arrived on deck, "armed with axes, belaying pins, cleavers, clinker bars and in fact everything movable that was on board. The pirates on seeing this array gave up the ghost and took to their boats."[47]

The analysis of letters composed by Enniskillen Township's foreign drillers suggests that, as the drillers experienced non-European culture for the first time, they reproduced the familiar gestures of nineteenth-century fiction writers and colonial officials. They replicated the rhetorical tropes of colonial discourse identified by post-colonial theorists and literary critics. The tropes identified in the narratives of the foreign drillers include: a visual mastery of the landscape from the writers' privileged point of view; an obsession with bodies, particularly the bodies of native women, accompanied with warnings about their seductive allure; appropriation of the territory by inclusion and domestication; classification and ordering of natural beings; idealization of the colonialist enterprise against the setting of emptiness and disorder; the notion of the "white man's burden" as a metaphor for the civilizing mission; and insubstantiation, or seeing the Orient as a dream.[48]

ENCOUNTERS IN THE "CONTACT ZONE" AND THE RE-INVENTION OF DOMESTICITY

During the nineteenth and early twentieth century, the colonial oil fields became a theatre for the re-invention of European imperialist ideals of domesticity. The domestic/colonial and civilized/backward relationship was never entirely, or homogeneously, either "home" or "empire." Nevertheless, it is a testament to the power of a common racial heritage that the white foreign drillers from Enniskillen Township became part of the imperial overclass. The drillers from Enniskillen Township enjoyed a position of privilege while working abroad by virtue of their race, gender, and technical skills in the extraction of oil resources. In the colonial oil fields the processes of racialization, class formation, and the construction of gender difference were all mutually constitutive.

Enniskillen drillers usually signed contracts of one, three, or five years in duration. Bill Gillespie signed a standard employee contract with Messrs M. Samuel & Co. of London, on behalf of the Nederlandsch Indische Industrie en Handel Maatschappu on 22 November 1904 for his services in Borneo. The company agreed to provide Gillespie with a second-class passage from Petrolia to Borneo. Gillespie was to work not less than eight hours per day once he arrived in Borneo. He was to be paid a rate of five American dollars per day, plus twenty-five dollars per month for the cost of food. Capitalist imperial authority was sanctioned using dominant middle-class ideals of morality and domesticity. The company reserved the right to stop the pay of any driller in the case of illness caused by "insobriety, immorality or other misconduct and to discharge the said employee summarily on any of these accounts." In the case of dismissal for behaviour deemed immoral by the company, the driller also forfeited his return passage to Petrolia and was legally bound to refund the cost of his passage from Petrolia to Borneo.[49]

A similar "morality clause" was included in the three-year contract signed by Ernest Kells on 1 March 1928 with The First Exploitation Company of England for his services in Persia. The contract also contained a clause that was tantamount to garnisheeing the driller's wages. The company was entitled to deduct monthly installments of £5 from his wages and pay interest at the rate of five per cent per annum on the sum retained. This was called a "Retention Fund." In the event of a driller terminating his contract before eighteen months had

Figure 5.1 Enniskillen drillers' bungalow in Miri, Borneo. Oil Museum of Canada, Oil Springs

passed, or if he was dismissed for insobriety or gross misconduct, the company was entitled to recoup its losses from the Retention Fund.[50]

The drillers' contracts also contained provisions for messing and servants. Gillespie's contract with M. Samuel & Co. stipulated that the employer would provide unfurnished lodgings, water, and light and that "the said employee shall provide his own native servant and pay for his own washing the said employers providing suitable unfurnished lodgings for the said native servant." Thus class differences between the white drillers as masters and their native servants were racialized and given capitalist imperialist sanction in the contracts signed by Enniskillen drillers with the oil companies. The same provisions were written into another subsequent contract signed by Gillespie with the Anglo-Saxon Petroleum Company on 12 July 1920 whereby he agreed to go to Sarawak for three years.[51] Similar provisions were probably included in the contract signed by Edward Winnett when he went to Sumatra in the fall of 1897. Winnett wrote in his diary on 16 December 1897 that he had purchased two wardrobes, two bedsteads, 12 chairs, a looking glass, mattresses, pillows, mosquito netting, and hanging lamps in Palambang. Upon returning to his bungalow the following day he found some natives fixing up the room and erecting a bathroom for the drillers.[52]

In his travel chronicle A.B.C. presents a view of master-servant social relations in which Victorian constructions of domesticity are naturalized. He rounded out his narrative of adventure in the "wilds of Sumatra" with a detailed description of their housekeeping arrangements and the "civilizing" influences of the Enniskillen drillers. This discourse was part of the process by which European imperialist domestic ideals were introduced and white Anglo-Canadian drillers were made masters of the non-white "savages." A Chinese cook was secured from Berandon. An arrangement was made whereby the cook would live with the drillers. "We were to pay him so much a month," he wrote, "and he would have to furnish everything to eat, and that saved us a lot of trouble, for we had to send to Berandon for everything."[53] The driller praised the native cook, indicating that he did very well despite that fact that he had no stove and only an open fire with which to prepare meals. The drillers also planted a vegetable garden. Tending the garden in Sumatra was probably the work of the native house servant.

The labour of non-white native Others was an important "civilizing" influence in colonial discourse, and another means by which class and racial differences were mutually constituted. During the year he spent in Sumatra, Edward Winnett supervised large crews of Chinese and Malay labourers hired to construct a pipeline and to unload metal boiler plate from steamships to be used in the construction of a new oil refinery. Conflicts sometimes occurred between the native workers and their white employers. On 12 September 1898 Winnett wrote [reproduced here exactly as written]:

> This is a very busy morning all the Coolies having struck for more money all together there was 360 men on strike they were all collected around my Bungalow they were a dangerous looking lot nearly every one of them had a cleaver or what they call a prong with them and others with large knives in their belts so we had to give them a raise of pay instead of lowering their pay so before noon we had them all satisfied and back to work.[54]

In his text Winnett defines the native workers as unruly and dangerous, although it would appear that wage reductions were imminent and most likely prompted the strike.

Subsequent Enniskillen drillers categorized the native peoples of Borneo and Sumatra using racial hierarchies which linked skin colour and physical attributes with moral characterizations and usefulness to the drillers in the oil fields.[55] "There are many tribes of the human race in Borneo, but the wildest most isolated, and never seen by the ordinary travellers, and sometimes by people working a number of years in Borneo, are the Punans, Murits and Land Dayaks, or Dyaks," one Petrolia driller wrote. These peoples, the driller revealed, lived in the jungle and lived off wild fruit and the roots of bamboo, and the meat of snakes, monkeys, and wild pigs. He described the Punans as "a strong, healthy, well-knit looking lot, light in color in comparison with other native tribes, probably from living in the dark recesses of the jungle." The Land Dyaks, who had been persecuted and hunted by the Sea Dyaks, lived in the mountains. According to the driller, the Land Dyaks were the most civilized of Borneo's native peoples. They were industrious and honest, and occupied villages of a permanent character. The Kengar, Kayans, and Sea Dyaks were elaborately tattooed from the hips to the knees. "The Kayans are supposed to be on a higher plane of civilization than the Sea Dyaks, though more cruel and their women are more inhumane than the men," the driller pointed out. The Kayans also had light skin. The Sea Dyaks were head-hunters who used the heads for religious purposes. In his travel chronicle the driller indicated that the Sea Dyaks inhabited the coastline of the island and were employed by the different oil companies for clearing the jungle for surveyors and geologists. They were also used as watchmen in the oilfields. Although they appeared to be peaceful people, the driller noted that "they had not lost the art of taking heads."[56] Another driller, writing two years earlier in 1923, indicated that the Dyaks "are absolutely worthless as workers but are unexcelled as watchmen ... and when put on a job will stay 'put'." All the heavy work was done by Chinese labourers.[57] This racial hierarchy further served to bolster the authority of the white drillers in the colonial oil fields.

The foreign drillers from Enniskillen Township helped to create a new sense of "whiteness" in the colonial oil fields by creating a hierarchy of racial difference that defined native peoples as non-white Others.[58] The lighter-skinned natives were always judged more useful to the white drillers. Dark-skinned natives were defined as "savages" who were untrustworthy and inhabited the jungles. Class formation and a systematic sense of whiteness developed hand in hand in the colonies of oil exploitation. The white drillers' technical skill and status of racial and class privilege afforded them positions of authority and prestige that eluded them at home. Bill Gillespie described for his neighbours back home how he, "just an ordinary driller," was given the privilege of meeting the Prince of Wales in Brunei, Borneo, in May 1922.[59]

Figure 5.2 Cementing a well in Miri. Oil Museum of Canada, Oil Springs

Enniskillen drillers, by virtue of their technical expertise in oil drilling and their developing sense of "Canadianness," defined themselves as more suited to the task of empire-building than Europeans. A few months after he started work in British Burma in the spring of 1880, John W. Crosbie penned a letter to John Clinton of Petrolia. In his letter Crosbie declared that the climate, "although disagreeable and frequently fatal to Englishmen, is not particularly hard on Canadians, as it resembles very much our July and August weather all the time."[60] The Canadian drillers who went to British Burma and to the East Indies supervised gangs of native labourers. The labour done by the "coolies," Crosbie reported, was not worth much, "one good Canadian being able to do more work than half a dozen of them."[61]

Enniskillen drillers who went to the East Indies learned the Malayan language and made it their own. In December 1899 M.J. Kelly was able to report that "Baldy" Stokes "talks Malayan like a native." The foreign drillers learned native languages to communicate with gangs of native labourers under their supervision. Robert Laird frequently lapsed into Malay when recalling his experiences in Borneo for the *Toronto Star Weekly* many years later.[62]

Figure 5.3 New Year's Eve Celebration at the company club in Miri, Borneo, 1924. Oil Museum of Canada, Oil Springs

Until after World War I, wives and children rarely accompanied their husbands to the colonial oil fields. Although it is likely that at least a small minority of the foreign drillers had homosexual relationships or sexual relationships with native women, these liaisons were not mentioned in the drillers' travel writings. In her study of the plantations of North Sumatra, Ann Laura Stoler found that plantation owners encouraged concubinage among native women and white overseers.[63] There was, however, no evidence of such practices in the colonial oil fields of the Dutch East Indies.

Some of the drilling companies constructed exclusive "whites-only" clubhouses for the amusement of their employees. Old Benwell went to the company clubhouse shortly after his arrival in Berandan, Sumatra, in the fall of 1898. The facility was equipped with a piano, billiard tables, a bowling alley, and all manner of games. At some drilling sites, however, the drillers apparently had to devise their own amusements. In a letter to the editor of the *Topic,* James McLean described a celebration organized by the "boys" to mark the completion of a new bungalow they had constructed. A band organized by the driller and dubbed the "Sanga Sanga band," provided the entertainment. "We had dancing as well," McLean wrote, "but unfortunately it was only a 'stag dance' for the want of the fair sex, which loss was greatly felt through the evening." Commenting on the lack of opportunity for contact with white European women in Sumatra, McLean remarked: "We should have given everything in our power to have had a few of our damsels to have graced the entertainment."[64]

A handful of women did accompany their husbands to the East Indies before World War I. J.W. Crosbie's wife spent fifteen months with her husband in Sumatra, returning to Petrolia in the spring of 1896.[65] In the early years of the twentieth century Robert Laird's wife accompanied her husband to Borneo. The Lairds spent two years at Sangiramik, where Robert Laird worked to develop the oil field. "During all that time," Laird stated, "my wife never saw a white woman." The Lairds were described in the pages of the *Toronto Weekly Star* as "Petrolia Orientalists." Mrs Laird praised the Dutch colonial cuisine. "We had some great banquets when we went to Batavia on our holidays," she declared. "I've seen as many as ten boys standing at the table waiting their turn to serve us."[66] Like their husbands, the wives of foreign drillers helped to promote the project of European imperialism.

The presence of more white women in colonial drilling communities after World War I put new demands on the oil companies to mark out their social space and to extend their measure of surveillance to ensure that white women were safe and native labourers were kept "under control." At Miri, Sarawak, in July 1923 the Chinese broke out in rebellion, although according to Petrolia driller Bill Gillespie, no one knew exactly why they rebelled or what they intended to do. When it was over twelve Chinese labourers were dead, and several more died in hospital. Two Europeans, both state officials, were also hurt.[67] In the aftermath of the Miri riot, the general manager of Sarawak Oilfields Limited issued a memorandum instructing its European employees "as far as possible to be impartial in their dealing with Members of the Staff and Labour Force of the various Nationalities." In the event of further rioting, the married Europeans were instructed to remain with their wives and families while the bachelors were ordered to gather together at one of the bungalows.[68] In the oilfields of Borneo and Peru during the 1920s separate white enclaves of company housing were constructed to protect the wives and children of foreign drillers.

Although women occupied a marginal position in oil production, they were the bearers of elevated moral standards in these insulated and isolated social spaces. In the 1920s oil companies and colonial administrators were rethinking the ways in which their authority should be expressed. The creation of European-like environments and the presence of white women were intended to insulate the men from the contamination of sexual contact with native women.[69] An Old Petrolia Boy, writing from Borneo in 1922 reported that "Miri has got to be a miniature of what Petrolia was in the old days," during the oil boom of the late nineteenth century. The white drillers' community in Miri had three clubs, two of them with billiard tables and dancing floors, and a church. The company had the latest "picture shows" brought in for the entertainment of the drillers and their families.[70] At least three Petrolia men who departed as bachelors married English women and came home with families. The oil companies employed a few single women as stenographers and nurses. In a letter dated 19 November 1920, and subsequently published in the *Advertiser-Topic*, William Gillespie informed those at home of "A Happy Event in Borneo." A.E. Haley, one of the Petrolia Boys, had married Miss Stonehouse, of Hull, England, in the Court House at Miri.

The drillers especially lamented the departure from Miri of one Petrolia woman, Mrs John Blake, in July 1921. Mrs Blake organized Christmas dinners where she gathered all of the single men.[71] As part of the project of reinventing domesticity in the colonies of oil exploitation, Enniskillen drillers and their families celebrated Western Christian holidays. One of the Boys recounted the Christmas celebration held in Miri in 1922 in a letter to the editor of the *Advertiser-Topic*. A Christmas tree was decorated for the children and all the latest toys were imported from England for the youngsters. "Christmas Day was much as it is at home, but without the cold and snow," the driller wrote. "Church in the afternoon, dancing early in the evening at the Rest House and then everyone off to the various dinner parties, music, song and story, until the wee small hours of the morning."[72] Similarly, Vola Braybrook, who accompanied her husband to Venezuela, recalled that "each family would take in a bachelor or two over the Christmas and Easter holidays."[73]

The households maintained by the wives of foreign drillers became sites for the reinvention of European ideals of domesticity. Class formation and the process of racialization were also mutually constitutive among the wives of foreign drillers. In 1921 Agnes Miller accompanied her husband Orville "Dick" Miller to Venezuela. "They spoke Spanish and I couldn't speak a word of Spanish at all," Miller remembered. "Anyway we took lessons in Spanish from a man who knew

Spanish and spoke good English." During the time they lived in Venezuela the Millers had two maids: "one was a cook and one cleaned the house. Polished the shoes." They also had a male servant named Dabro, who washed Dick Miller's car. Agnes Miller indicated that having servants "was part of the game." Servants "were very cheap in those days," earning ten dollars per month. She also described the native servants as "thievish."

The Miller's enjoyed a class status and a standard of living above that of the families of drillers who remained in Enniskillen Township. Dick Miller had his own car. He earned U.S. $225 a month, twice the amount he would have made back home, from his job as a tool dress, where he put the final finishing on the drilling tools, shaped the drill bits, and repaired used and damaged drilling equipment in the field. He was soon made a driller, when he earned even more. Agnes Miller was relieved of many of the housekeeping duties required of the wives of drillers who remained at home. With two maids doing the bulk of the housekeeping, Agnes Miller was free to spend afternoons playing bridge with other drillers' wives. The Millers spent their evenings at the company club. The cook would have dinner prepared for them when they returned.[74]

Vola Braybrook recalled some of the difficulties she experienced in reinventing domesticity while living abroad: "You see they didn't think anything of cleaning a milk pitcher in the kitchen and taking the rag and wringing it out and cleaning all the hand basins and the bathroom equipment." Cleanliness was at the heart of the Canadian women's ideal of domesticity in the contact zone. Although the voice of the native women domestics was obviously excluded from the oral histories provided by the wives of Enniskillen's foreign drillers, their accounts suggest that the experience was not perceived by the native women in the same way. Vola Braybrook noted that when a "local girl" came into the household she was usually frightened.[75]

The racialization of class difference among the families of foreign drillers and their native servants was not always fraught with tension and conflict. Dorothy Stevenson, who went to Venezuela when she was only eight years old, claimed that her mother "took the maid into the family." The maid shared meals with Dorothy and her mother. "She didn't eat with us at night when my Dad was there, but all the rest of the time and we were a family," Dorothy recalled. Perhaps in some instances gender muted, but did not erase, relations of power constructed on racial and class lines, which were central to the invention of imperialism; and possibly racial lines were less clear in South America than they were in the colonial oil fields of the East Indies. Dorothy Stevenson suggested that the rural working-class background of the drillers encouraged harmonious relations with the natives – "they were normal farm boys ... I know Canadians were well liked."[76]

The children of Enniskillen's foreign drillers usually remained with their parents in foreign fields until they were of school age, whereupon they returned home. Children were believed to be particularly impressionable; thus they were sent back to Canada to avoid the "danger," and cultural contagion, that was thought to arise from the prolonged interaction of white children with natives. Most of the children were accompanied by their mothers and attended local schools. In the fall of 1923 one Petrolia teacher remarked that during the previous year she had taught children who have lived in every continent but Australia. "A geography lesson was full of interest and life," she remarked.[77]

In October 1923 the drilling community established by the International Petroleum Company at Negritos, Peru, reportedly contained 8,146 Peruvians and 192 "foreigners." The "foreigners" comprised 90 men, 51 women, and 51 children. The company provided racially segregated schooling organized into three units: white, Peruvian, and Anglo-Peruvian. A separate school was provided for white European and North American children, commencing with primary grades and finishing with third form at the secondary-school level. The teachers were engaged in the U.S. The Peruvian school was under the direction of a Peruvian superintendent. Children of "mixed parentage" – one Peruvian parent and one English, Danish, or Spanish parent – attended the Anglo-Peruvian school. The company provided the books and supplies for the school children.[78] Agnes Miller returned to Petrolia shortly after her daughter Dorothy started school. "She wasn't going to get very much Canadian education in Colombia," Miller stated.[79]

Enniskillen's foreign drillers enjoyed class and racial privileges in the colonies of oil exploitation. This was sustained not only through formal means such as the provision of company housing, schools, and whites-only social clubs but also in more informal ways, including the practice of greeting fellow Enniskillen drillers at the dock as they arrived in foreign fields. So many Petrolians lived in Borneo in the early 1920s that Mrs Alfred Webb, in a rare letter from the wife of a foreign driller published in the local newspaper, commented that "it does not seem that we are really in a foreign country."[80]

DOMESTICITY AND DISPLAYS OF IMPERIAL SPECTACLE AT HOME

According to Vola Braybrook, the wives of Enniskillen's foreign drillers did not really want to go abroad: "They wanted to raise the children among the relatives and in Canada," she stated.[81] Arnold Thompson's father drilled in Egypt, Cuba, and India during the period surrounding World War I. Thompson described the impact that the long periods of separation had on familial relationships:

> when I was six years old my father left for foreign fields. From then on my father was a stranger. People will find it difficult to accept that. But this is a fact of family life, not just with my family, but with a number of families in Petrolia ... So therefore our mothers brought us up ... When our father came home ... I'm going to speak for myself. When my father come [sic] home he was just another man. I didn't know him. I didn't know what to talk to him about. He didn't know what to talk to me about because we had nothing in common.[82]

It was not unusual for the children of Enniskillen's foreign drillers to report that their fathers were complete strangers when they returned home after an absence of three or more years. Arnold Thompson also revealed that relations between husbands and wives were oftentimes strained after a separation of several years.

In their reminiscences the children of foreign drillers remember their mothers as being extremely busy and lonely. Many wives relied on extended kinship networks for support during the long periods when their husbands were absent. Dorothy Stevenson recalled that her father would no sooner arrive home than he would announce that he had signed another contract to drill abroad. "And then mother would pack up her furniture and go home to her mother," Stevenson stated. Her grandfather was the "man in the house," and he became a "surrogate father" to Dorothy.[83]

Many of the wives of foreign drillers were active in local church groups and attended services every Sunday. Ev Brimmer revealed that her mother used to invite the Ladies' Aid from the Presbyterian Church to their home for afternoon tea while her father was absent abroad.[84] The wives of foreign drillers also organized informal bridge groups and held quilting bees. Extra-marital sexual relations were allegedly rare, although many of those interviewed were children at

Figure 5.4 Petrolia drillers in Singapore. Oil Museum of Canada, Oil Springs

the time and thus were unlikely to be aware of any sexual liaisons involving their mothers. "You know women lived a very restricted life," Dorothy Stevenson said. "They just weren't free to wander around as we do now."[85]

The homes of the foreign drillers in Petrolia and surrounding Enniskillen Township became a space for the display of imperial spectacle. The drillers mailed home postcards made in photographic studios. "Photography became the servant of imperial progress," historian Anne McClintock suggests. In the colonial postcard, "time is reorganized as spectacle; history is organized into a single linear narrative of progress."[86] The postcard "Petrolia Drillers in Singapore," was staged in a photographic studio, and was probably paid for by the men sitting in the rickshaw: Bill McCrae on the left and Bill McEwen on the right. Imperial authority is elaborately staged in the photograph. The drillers are displayed in a position of prominence and authority in the white summer dress of a gentleman in a rickshaw drawn by an Oriental boy. In contrast to the finery of the two drillers, the boy wears a rough straw hat and is bare chested.

The drillers also purchased curios and had them shipped back home. As Ev Brimmer recalled:

> We would get funny things. It would seem to a kid to be funny things ... One I used to see on the wall Mother would tell me was a sword fish's tongue. I don't know whether it was a tongue. It had horrible teeth all over it. It was hanging on the wall. There was a war club from Persia strapped on the wall. And then one day I walked in the living room. We called it the parlor then. And on the floor were two of the most beautiful Persian rugs. Just absolutely gorgeous.[87]

The curiosities, the postcards, and the travel chronicles all added to the prestige of the foreign drillers in the community.

In addition to the position of authority the drillers attained in the community as travelled men, they also earned high wages. "We weren't short of anything you know," Ev Brimmer stated, "we were I guess well off":

> Persian rugs, cherry whatnots, cherry wardrobes. We had beautiful things. A player piano. The newest thing that came out we would have it. Player piano, Persian rugs ... Limoge [sic] china that we ate off every day. Limoge china. I remember breaking the gravy boat, knocking the gravy boat off the platter. And Momma, just ordinary. Just ordinary. Limoge. I would give my eye teeth for some of that china we ate off now. And we just took it for granted.[88]

Most drillers made arrangements to send money home regularly every month. Dorothy Stevenson recalled that her father deposited "house money" into an account set up at the Petrolia branch of the Bank of Toronto.[89] The wives of the foreign drillers managed expenditures for household consumption in the absence of their husbands.

The local economy and imperial capitalist expansion were interwoven in the process of class formation. As another crew of Enniskillen drillers set off for Borneo in May 1899, the *Advertiser* commented:

"Our men go abroad unloosing the riches of the earth in strange lands, but they almost all come back to Petrolia to enjoy the rich fruits of their labor, much to the advantage of our town." Twenty-five years later, in a letter signed One of the Old Timers, a retired driller extolled the economic benefits to the community that resulted from the employment of local drillers abroad. The Old Timer revealed that "$15,000 or more a month comes back to Petrolia from these drillers, who exile themselves for years on end, that they may have a 'stake' for their old age."[90]

While it was not possible to trace the career path of all of Enniskillen Township's foreign drillers, it would appear that employment abroad resulted in elevation into the ranks of the middle class for a substantial number of foreign drillers. A handful of Enniskillen's foreign drillers became wealthy men and, like William McGarvey, never returned to the community. By the 1920s some of the drillers with experience in foreign fields, who had learned drilling at what was by then referred to locally as the "Petrolia Drillers University," attained supervisory and management positions with large international oil companies. George W. Brake learned the trade in the oil fields of Petrolia from his father and grandfather, both of whom were oil drillers. In 1902 Brake went to the Dutch East Indies and remained there until 1905, when he signed a contract with a syndicate to drill in Egypt. When World War I broke out Brake was drilling in Venezuela. During the war he relocated to Peru, where he became superintendent of the production department. Another local driller, Joseph Burns of Oil Springs, was superintendent of the drilling department with the International Petroleum Company in Peru. In 1922 Fred Sander was made the general superintendent of a refinery operated by the Royal Dutch Oil Company in Tampico, Mexico.[91] Several foreign drillers spent most of their adult working lives abroad and managed to save enough money to retire to Petrolia or Sarnia. Some foreign drillers were able to purchase farms, start small businesses, or purchase their own local oil properties. Others went to work locally at the Canadian Oil Refinery in Petrolia or relocated to Sarnia where they found employment. For instance, John Keene drilled for oil and water in Egypt, Borneo, Sumatra, Venezuela, and at various points in Canada during the early decades of the twentieth century. Upon returning home, he was hired as an engineer in the Petrolia fields. He then went to work at the Petrolia Wagon Works before securing a position in the Canadian Oil Refinery, from which he retired in 1943.[92] A considerable proportion of Enniskillen Township's foreign drillers rejoined the ranks of the working class upon returning home. Dorothy Stevenson noted that the men who were unable to save any of their wages while working abroad "had a problem when they came home, and they had to work for somebody else."[93]

Local manufacturers and retailers also benefitted from the imperial project in colonial exploitation. The Oil Well Supply Company, which still exists today, provided drilling tools and rigs for colonial oil exploitation from its earliest stages.[94] In January 1900, when the local economy was in desperate straits in the aftermath of the relocation of the Imperial Oil Refinery to Sarnia and the decline in production in the local oil fields, the Oil Well Supply Company enlarged its blacksmith and machine shop plant in Petrolia to meet the demands of the foreign drilling trade.[95] This, in turn, provided employment for a highly specialized and skilled body of local workers. The local economy and global imperialist capitalist expansion were mutually dependent on one another.

As one passes through the present-day town of Petrolia, the legacy of the foreign drillers can be found in some of the street names derived from the foreign fields where Enniskillen drillers were employed, namely Oozloffsky, Ignatieff, and Valentia. Petrolians are proud of their heritage as "The Town of World Travelers," and have used this heritage to build a community identity.

CHAPTER SIX

"Canada's Victorian Oil Town": History, Public Memory, and Community

In the summer of 1898 the U.S. conglomerate Standard Oil acquired control of the Imperial Oil Company and operations were transferred from Petrolia to Sarnia. Petrolia was no longer the headquarters of the Canadian oil industry. During the early decades of the twentieth century attempts at industrial reorientation and diversification were made in Petrolia with varying degrees of success. Another oil refinery, independent of Standard Oil, the Canadian Oil Refinery Company, was chartered on 1 August 1901. Nevertheless, following the departure of Imperial Oil and the decline in local crude resources, the "boom town" *mentalité* that had characterized the community since the early 1860s dissipated. The exodus of skilled male drillers to foreign fields intensified. Over the course of the twentieth century numerous attempts were made by civic-minded Petrolians to reassert community identity by forging connections between memory and place through public spectacles and festivals. Beginning in the 1970s, as in other communities throughout North America and Western Europe that faced de-industrialization, government-sponsored initiatives in Petrolia promoted heritage and tourism with the intent of revitalizing the stagnant local economy.

Building public memory was part of the process of re-imagining community in Petrolia. History, memory, and an awareness of place have been central to the creation of "imagined communities." As Jenéa Tallentire writes, "The construction of communities is a social process, reproduced in the interactions of social networks, and represented by signs and symbols in the imaginings of individuals internal and external to the community."[1] Historians have taken Benedict Anderson's notion of "nation" as an "imagined political community" and applied it to various types of social groups with a shared identity and history.[2] In recent years the research focusing on the importance of public memory to the articulation of "imagined community" has been steadily increasing. A community's values, beliefs, and practices are expressed through the creation of particular narratives about the past, which are essential to its identity and cohesion.[3]

Shared sites of individual memory are essential to the formation of public memory and the imagining of community. As Raphael Samuel writes, memory is not merely a passive receptacle, rather it is active, dynamic, and historically conditioned. "It is a social form of knowledge," he suggests, "the work, in any given instance, of a thousand different hands."[4] Memory begins when something in the present stimulates an association with the past. In other words, people reshape their recollections of the past to suit their present needs.[5] Memories are not solitary constructions, as Maurice Halbwachs reminds us: individuals shape, omit, distort, recall, and reorganize their memories in interaction with others. Sharing, discussing, and interpreting individual memories develops collective and public memory. What becomes important in the historical study of memory is not how accurately recollections can be located in some piece of a past reality, but why historical actors construct their memories in a particular way at a particular time.[6]

The importance of the connections between memory, performance, and place to the imagining of community in Petrolia is explored in this chapter using the three Old Boys' reunions held in 1908, 1925, and 1946. These community celebrations exposed the political nature of history and public memory, as well as the connections between individual and public memory. A concern with "the power of place" in nurturing memory is particularly useful for considering how public memory was constructed in Petrolia around the Old Boys' reunions. Ordinary landscapes – towns, neighbourhoods, streets, and buildings – nurture public memory by encompassing shared time in the form of shared space. According to Delores Hayden, "places trigger memories for insiders, who have shared a common past, and at the same time places often can represent shared pasts to outsiders who might be interested in knowing about them in the present."[7] People's attachments to places are material, social, and imaginative. The Petrolia Old Boys' celebrations were effective in claiming the symbolic importance of place in shaping public memory.

Figure 6.1 Old Boys' Parade, 1925. Lambton Room, Lambton County Library, Wyoming, Ontario

Petrolians concerned themselves with promoting heritage as tourism in an attempt at economic redevelopment. This has been a political as well as a social and cultural process, involving local politicians, community volunteers, and various federal and provincial government local incentive programs. Saving a public past in Petrolia has involved decisions about what to remember and what to protect. This work has involved architects, collectors, local history buffs, civic officials, and community volunteers. At times "authenticity" has been compromised by other pragmatic and market-driven imperatives. The cultural displays that constitute the memory of a society are diverse and incorporate markers of place such as photographs and popularly understood images.[8] The symbolic and material representation of Petrolia as "Canada's Victorian Oil Town" was a concerted strategy to represent place simultaneously for the tourist gaze and the self gaze of the community. The heritage/tourism initiatives in Petrolia were contested political and ideological terrain. The reorientation of Petrolia's economy around heritage and tourism has been accomplished through the ongoing construction of a living museum called the Petrolia Discovery; the restoration of Victoria Hall, built in 1889, as a combined town hall and opera house and presently the site of a successful summer theatre program; and the preservation of the downtown façade.

THE OLD BOYS' (AND OLD GIRLS') CELEBRATIONS OF 1908, 1925, AND 1946

Old Boys' reunions were organized in Petrolia in 1908, 1925, and 1946. The first reunion, in 1908, was organized in the midst of the industrial and commercial reorientation that occurred in the aftermath of the removal of the Imperial Oil Company to Sarnia and the exodus of skilled drillers to foreign fields. The reunion was organized in the hope of reconstituting community. In the aftermath of the two World Wars, Old Boys' reunions were organized to bring back foreign drillers and their families and war veterans who settled elsewhere after the wars were over. During the planning and staging of the three Old Boys' reunions public memory was created or recreated. On the surface these celebrations were stereotypical small-town festivals, in which the citizens of Petrolia celebrated themselves in a community-wide expression of fellowship. Public memory, however, does not simply emerge out of the collective unconsciousness of a single-voiced organic community. The public memory created by these community festivals was a contested culture where class and gender differences, in particular, were evident; a field of both political and cultural forces. The social ties in communities are fragile. Nevertheless, as Robert Lavenda suggests, "In small towns, these ties must be nurtured and protected, for divisiveness is a danger to the community's survival as a more or less coherent entity."[9] Small-town festivals and celebrations are highly effective in claiming the symbolic importance of place. In the case of Petrolia, the building of a community identity was plausible and necessary because of its size.

In the fall of 1907 Alderman W.J. Clarke asked the town council to take up the issue of organizing an Old Boys' reunion in Petrolia and a public meeting was called for October 3. Approximately one hundred men attended the meeting. The outcome was the organization of the "Petrolia Old Boys' Association." A slate of officers, representing the town's male, professional, and business elite, was chosen. John H. Fairbank, "the Father of Petrolia," was selected honorary president, and James Peat, a drilling contractor, was chosen as president. Several standing committees of middle-class men were organized to complete the arrangements for the reunion scheduled for 13–14 August 1908. No women or working-class men were appointed to any of the committees; they would have different access to the community of celebration. Following the example of Forest, another small Lambton County community to the north, Petrolia set a membership fee of one dollar for resident and non-resident Old Boys. A lesser fee of fifty cents was set for Old Girls.[10] The Printing Committee prepared a circular to advertise the event. Former residents organized Petrolia Old Boys' Associations in Winnipeg, Toronto, Brantford, and St Thomas.[11]

The festival text in the pages of the two local newspapers emphasized camaraderie. "The purpose of the reunion is not for monetary gain, but that every individual in attendance shall have a grand good time – the time of their lives, in fact – where present and former residents of Petrolia and all the surrounding county – will be privileged to meet again and renew old acquaintanceships, and recall for a brief season the happy days of girlhood and boyhood," the *Advertiser* reported.[12] Yet the community was also imagined as a progressive commercial and industrial centre by the phalanx of middle-class businessmen, newspaper editors, and leading citizens who organized the event. Petrolia's businessmen saw the festival primarily as civic boosterism. They constructed progressive images of their community for the realization of practical economic and social benefits.

The day before the celebrations were set to begin, the *Topic* ran an advertisement for Stirrett's dry goods store. Robert Stirrett was a member of the town council and the Board of Trade. He also sat on the Finance Committee organized for the reunion. Below a large portrait of Stirrett was the by-line, "Should auld acquaintance be forgot." The remainder of the advertisement reads.

> We cut out all hints of business in our efforts to welcome the "Old Boys." One of the pleasures of a business life is the making of many friendships; one of the best triumphs of a business life is the retaining of many friendships, and one of the happiest incidents of a business life is the renewing of friendships. We are hoping for the glad hand from every "Old Boy." We wish you a pleasant reunion in the old town, and a safe return.
>
> You will find a waiting-room fitted up in the rear of our store for the convenience of visitors. We present you the freedom of the store.[13]

Hardware merchants VanTuyl and Fairbank invited "Old Boys and Girls" to inspect their "large stock of useful articles suitable for a gift to yourself or a friend unable to come."[14] Local businessmen clearly hoped to benefit financially from the festival.

The Old Boys' Association prepared a souvenir booklet, which sold for 25 cents at most business and mercantile establishments in the town. The booklet entitled "Petrolia, Canada 1862–1908, Its Advantages as a Commercial and Residential Centre," presented an image of the town closely aligned with the economic interests and civic-minded morality of the event's male middle-class organizers. This civic boosterism, however, played on an idea of local history. In a highly romanticized "Sketch of Petrolia's Early Days," John H. Fairbank told the story of the building of Petrolia and the "oil boom." Fairbank used warrior metaphors to commemorate the oil producers, who "early in the fight," during the mid-1860s, built the town. Fairbank's narrative focuses on public institutions, and the commercial and industrial establishments that the "founding fathers" built in the community. There are no references to the role of women in the public sphere in Fairbank's use of local history. Contrary to the representation of "the wild west" resource town where lawlessness and debauchery prevailed, Fairbank emphasized that, "Petrolia has always been an orderly, law-abiding, Sunday-observing, church-going place," with representatives in every land who look back with pride and pleasure to the "Old Town." Dominant Victorian middle-class family values and ideals of separate spheres were used in the essay. He continued, "At home and abroad, its men have been manly men, its women, womenly women, and its babies, perfectly lovely and numerous – at times, two on a stem. The baby carriage always has the right-of-way in Petrolia." Through this constitutive narrative the moral and social values of the event's planners were presented as those of the community and they created a public memory of Petrolia as a respectable Victorian oil town.

Several photographs were included in the 1908 souvenir booklet as visual markers of place. The middle-class male organizers of the Old Boys' reunion produced an aesthetic Petrolia through these visual representations. A collage of Crescent Park depicted the large, elaborate residences of the middle-class elite, who lived on "Quality Hill." At the centre of the collage is a photograph of middle-class men and women playing tennis in the park. Panoramic landscapes illustrate the transformation of Main Street from a collection of wooden shacks in 1866 to modern brick blocks in 1908, thus capturing the "boomtown" atmosphere that the organizers of the reunion wanted to convey. These and other highly sanitized representations of the town's business and public buildings helped to reinforce the privileged status of wealthy citizens like J.H. Fairbank, the Kerr brothers, George Moncrieff, and Jacob Englehart in the community. The working-class homes and businesses in the "Pithole" and the Blind Line were displaced by this bourgeois view of landscape. Still the members of the town's working class could probably identify with the photographs of leading industrial establishments, although they too were pristine in the representations, including the Canadian Oil Refinery, the Petrolia Packing Company, the Lambton Creamery, and J. & J. Kerr's lumber yard. None of the grease and grime of an oil-resource town is evident in any of the photographs.

The 1908 Old Boys' celebration opened on August 12 with a public reception and registration at the town hall. The reception committee met all incoming trains, and extended a hearty welcome and the "freedom of the Town to all ex-Petroleaites and visitors." Later that morning the "Grand Parade" marched down Main Street. The street parade, as anthropologists and social historians have explained, was a setting for spectacles and symbolic displays. As Mary Ryan and Cecilia Morgan point out, parades and civic ceremonies provide occasions to ferret out the cultural meaning attributed to sexual differences. The 1908 Old Boys' parade displayed the public in deeply gendered imagery.[15] The procession was organized by men, conducted by men, and marked by masculine symbols and associational representation. Reportedly a mile long, the parade was led by representatives of the militia: Major John H. Wynn of London, an "Old Petrolea boy"; Charles O. Fairbank; and Capt. Charles McKittrick. The procession included the Windsor Band and Old Boys; the Petrolea Oddfellows' encampment; a carriage containing the Mayor and the town council; the water works department; the Knights of Pythias Band; the fire department; the 7th Battalion Band of London and one hundred Old Boys; the Detroit Band captained by George Lewis, a Petrolia Old Boy, and 125 Old Boys; the Brigden Band and Old Boys; the Forest Band; Callithumpians; and representatives of the industries of the town. While token references to "Old Girls" were made in the newspaper advertisements, women were primarily on the margins as spectators. The Old Boys, particularly those who returned from foreign fields in Persia, Borneo, West Africa, Australia, and Mexico, were the main attraction. The return of the foreign drillers to Petrolia symbolized the importance of the community globally, and suggests that a local/ global connection had been forged in the construction of community identity. Significantly, the farming community surrounding Petrolia, upon which the town also relied for its economic survival, was excluded from the procession.[16] In 1908 community identity was organized around Petrolia's local history as an oil resource town interwoven with the global renown of its foreign drillers.

Running races and baseball matches were held over the course of the three-day celebration. On the second day of the reunion an oil well, a symbol of Petrolia's origins as an oil-producing community, was torpedoed on the flats of Bear Creek. The event's organizing committee furnished a variety of attractions and amusements, including band concerts, an acrobatic exhibition, a midway, and moving picture shows. The intent was to promote the event as a family attraction.[17] While the small-town nature of social relations in Petrolia in 1908 was perhaps too intimate to imagine any serious social tensions and conflict, differences of class and gender were evident in the organization of the Old Boys' reunion. Nevertheless, the event was declared a success by its male middle-class male organizers.

In the summer of 1924, following the upheaval and long recovery in Ontario in the aftermath of World War I, some of the town's citizens clamoured for another reunion. A public meeting was held in the council chamber on August 22, where Mayor Drummond issued a proclamation calling for a public meeting on September 2 in the interest of an "Old Home Week" in 1925. The supporters of the event constructed a particular idea of place, where the town was represented as home. The event provides insight into how the construction of public memory depended on individual memory, particularly the childhood memories of residents and former residents. The *Advertiser-Topic* emphasized the connections between childhood memory, public memory, and community identity in its endorsement of the reunion: "No man ever gets so far away that he doesn't sometimes just yearn to be back among the old crowd of pals and talk over the times when they were kids together. To turn the clock back to those very same days of childhood and to give our busy town to the Old Boys and Girls for a day or two would be a great source of delight to the older residents of the town and to the many Petrolians who years ago departed from their native heath, but who still remember the old town, and still fondly repeat those lines 'How dear to my heart are the scenes of my childhood.'"[18]

Another public meeting on 2 September 1924 was poorly attended. While some of those in attendance decided to form a committee to organize an Old Home Week celebration to be held 1–5 July 1925, some expressed their reservations about such an event. Dr John Dunfield was pessimistic about the proposed reunion. He spoke of the considerable expense to the town of the 1908 celebration and indicated that he would "have nothing to do with it." Blacksmith James

Sanson also expressed his reservations. "We need more enthusiasm," Sanson remarked. "Let us have a celebration if we can get the necessary enthusiasm to put it over. We don't want to do things half way."[19] Despite the concerns voiced by some of the men attending the meeting, plans were made to hold another meeting the following evening.

This meeting was reportedly "well representative" of the citizens of the community, although once again the town's male middle class dominated the proceedings. James Peat, President of the 1908 Old Boys spoke first: "The last Old Boys was a remarkable affair and excelled any similar event in the country. Unless the citizens in general make up their minds to its success it will be an absolute failure. If any one is anticipating to make money out of it, from a personal point of view, they better forget it."[20] The outcome was the organization of the Petrolia Old Home Committee, composed entirely of middle-class professionals and businessmen, as was the 1908 committee.

Again a membership fee of $1 for adults and 50 cents for children was levied to cover expenses. Letters were sent out, and an invitation booklet was issued, urging "Old Boys and Old Girls" to "Plan Your Vacation To Get Here." Over the next several months the *Advertiser-Topic* promoted the event. Residents and former residents were encouraged to send their membership fees to the secretary of the organizing committee. The members' names were published in the local newspaper. W.P. Macdonald, secretary of the organizing committee, also sent letters to Old Boys and Old Girls who had moved elsewhere inviting them to "Come on Home."

In early November 1924 the *Advertiser-Topic* revealed some of the plans for the celebration:

> The early history of Petrolia oil town and the county of Lambton are to be depicted in fun, frolic and pageantry. Early pioneer life of Canada's oldest oil fields, together with the farm pioneer life will be impersonated in character, dress and by picture. A gala celebration calling forth the memory of by-gone days is the endeavor of the committees working strenuously to make the coming celebration the greatest ever attempted.[21]

In an important departure from the 1908 celebration, the event's planners decided to integrate the surrounding Enniskillen Township farm community into the spectacle.

Another difference from the 1908 celebration was the involvement of women. The local branch of the Council of Women formed a Ladies'

Reception Committee to procure a restroom for the ladies and to organize a reception area for visiting Old Girls. The council was also given the job of visiting every household in the town to arrange for the billeting of the visitors. In addition the women marketed souvenir badges to commemorate the occasion and organized a historical exhibit of "curious articles native and foreign to this country, relics, collections of any nature, articles of historical value or depicting past or present events in travel." As Beverly Boutilier and Alison Prentice have demonstrated, community-building was one of the prime motives of English-Canadian women who engaged in historical work.[22] The historical exhibit served to reinforce the representation of Petrolia as a well-travelled, cosmopolitan community built around the global renown of the town's foreign drillers.

The 1925 Old Home Reunion prompted the creation of public memory through a renewed interest in local history and an attachment to place. In a letter to the editor of the *Advertiser-Topic* published on 21 May 1925, under the heading "Old Town Relics," the correspondent, who signed his letter Old Boy, suggested that a collection and display of "things" relating to the early history of Petrolia might be organized in connection with the reunion. The anonymous Old Boy provided examples of what he had in mind. He suggested that the following be included in the exhibit: a copy of the first newspaper published in Petrolia, an invitation to the inauguration of the Oil Exchange Hall, in 1871, a Bill proclaiming Petrolia a town in 1874, and a program for the opening of Victoria Hall in 1889. Public memory was constructed around relics of the town's public places, most notably, the Oil Exchange Hall, a potent symbol of Petrolia's origins as an oil resource community.

Public memory was also constructed in 1925 through an attachment to Petrolia as "home," based on remembrances of childhood. On 25 June 1925, one week before the reunion, the *Advertiser-Topic* published a special souvenir "Petrolia Old Home Week" edition.[23] Former Old Boys, who had left Petrolia and "made good" elsewhere were asked to contribute their childhood memories. The editor referred to the role of the newspaper in keeping the project alive: facilitating the creation of a public memory by organizing individual perceptions of the past. "These events do not just happen," he wrote. "They have to be read, marked and inwardly digested, or in other words, discussed, re-discussed, and then discussed again, until the public is aroused to a point where many minds think alike." Memory is an active process; the heading "I Remember," appears above each narrative in the special edition of the *Advertiser-Topic.* Memories are supported by the recollections of others; they are public and shared.

The individual memories of those who wrote for the newspaper were a lived history of family, childhood playmates, and school chums. From the individual memories of the Old Boys, a public memory was constructed around the representation of the town as "home." "While born in England, I feel that Petrolia is my home town, and have always been interested in everything connected with it," remarked A.F. Rutter, vice-president of the publishing firm Warwick Bros & Rutter of Toronto. "Everybody was good to me as a youth, and it is the hallowed resting place of my parents," Rutter continued. "I would be ungrateful indeed to forget that it gave me a companion who has taken care of me all these years, as well as a family that is now in the second generation."

The Old Boys remembered their school days, boyish pranks, and first jobs. Dr J.F. McKitrick of Los Angeles wrote, "My earliest recollection dates back to August 1882, when I first went barefooted to Miss McRobie's school room, where I sat between George (Juddy) Glover and Jack Gleason. Very closely associated with my recollections of these school days is the remembrance of the cat-of-nine-tails. No doubt the memory also lingers in the minds of others." W.F. Reynolds remembered the evening that he and his friend Alex Watson "invaded" Russell's Wyoming farm, where they were caught stealing cabbages by the owner. W.G. Fraser gave A.F. Rutter his first job, and "perhaps would have kept me, had I been able to carry a bag of oats after filling it," he remarked.

Geography acts as a sort of "memory theatre": a direct aid for the recall of the past and a structure for remembrance. Childhood play areas allowed the Old Boys to locate their memories in time and space. J.W. Craise, branch manager for the Excelsior Life Insurance Co. in Regina remembered:

> A great pastime in summer was Woodley's swimming hole, and the shady spots in the evening on Main street west, always known as Lover's Lane. The winter evenings were spent at Cooley's rink, and later at Columbus rink, skating the light fantastic with my good old pals of both sex, sleigh ride parties to country homes, Wyoming and Oil Springs, and tobogganing on the Englehart hill in the moonlight.

Craise and the other Old Boys reminisced about their childhood using the places that were the sites of childhood amusements. Individual remembrances of these local childhood pleasure spots were part of the process through which public memory was spatially organized.

Family connections to the oil industry were included in the memories of childhood. F.W. Reynolds told of his parents selling their farm near Ingersoll in the 1860s, and "hastening to find the pot of black gold at the end of the rainbow." He recalled that "the most valuable well ever drilled in the local field" was struck by his father in partnership with Joseph McDougall, William H. McGarvey, and Neil Sinclair. "It was named 'The Deluge,' controlled with some difficulty, and yielded 400 barrels a day over a considerable period." Reynolds's most vivid, and humorous, childhood recollection attests to the pervasiveness of the oil industry in the process of constructing public memory. "Around the age of six or seven, I had a pair of red-top boots, with brass toes, and undertook to wade through the limestone at a new well on the hill back of the barn," Reynolds recalled. "Cyrus Perkins was the driller, and I can still hear his laughter as I slipped and rolled down hill in the sticky mess." Reynolds's experience was shared by many of the Old Boys and Old Girls raised in Petrolia.

The Old Boys suggested that childhood associations with Petrolia continued into their adult lives, even after they moved away, bridging their childhood past with their adult present. These affiliations were more remarkable in the case of Petrolia because of the town's status as a global provider of expert oil drillers. Charles Fraser recalled meeting Gordon Stirrett, an engineer, one Sunday afternoon while swimming in the Pacific Ocean near his home in Vancouver. Fraser wrote:

> I passed a husky looking chap who said to me, "I wish I could swim like that." I turned and took a look at him, for I thought he was making fun of my *avoirdupois.* I saw that he had that structural steel build himself and I said, "You can swim equally as well as I do if you only make up your mind to and strike out." "No," said he, "I learned to swim in a place called Schumann's where they didn't go in for long distance.' I took another look and said, "Friend, you have nothing on me, for I learned in the same old hole."

These tales of chance meetings in other localities with men and women who shared a connection with the town of Petrolia contributed to the building of a public memory out of a shared lived history.

As the July 1 beginning of Old Home Week approached, the organizing committee asked the citizens of the community to decorate their homes and places of business with flags and bunting in patriotic designs and colours. The Registration Committee issued a call to every resident and visiting Old Boy and Old Girl to register, so that a complete record could be kept as a historic remembrance. The three-pole oil derrick and pump, which covered the countryside during the heyday of the oil boom during the nineteenth century, was designated the official emblem of the 1925 Old Home Week celebration. Every schoolboy and schoolgirl was asked to build a replica on the front lawn of the family residence. Cash prizes were awarded for "the best erected miniature derrick." Builders and lumber merchants, J. & J. Kerr Co., constructed a 72 foot derrick on the main street in front of the railway station, complete with electric lights to illuminate the structure at night.[24]

The first day of the celebrations coincided with Dominion Day. Billy Stewart, of London, Ontario, was chosen as the representative Old Boy. During the opening ceremonies he was presented with a symbolic key to the town and a lantern "in case the old boys and girls were out late." Stewart addressed the crowd referring to old memories and scenes of childhood, which he claimed, "could never be erased from his heart." Immediately after the presentation, the parade commenced. As in the 1908 celebration, the parade was led by the "town fathers." Next came a float with five little girls on top symbolizing a citizenry that required the protection of the robust male public officials who led the parade. Scottish Highland "Kiltie" bands from London and Sarnia, floats representing local industries, Petrolia's servicemen, the Boy Scouts, Callithumpians, and a kazoo band followed. At midday the Windsor contingent of Old Boys arrived in sixty-five rail cars. They were met at the east end by the Reception Committee and escorted into town by the band from Bridgen. Sports events took place in the afternoon, and the Harry Lotteridge Amusement Company erected its midway around Victoria Park and the town hall.

The second day of Old Home Week was designated "Border Cities' Day," in honour of the large group from Windsor that attended the event. The Border Cities Old Boys assembled at their headquarters in Lancey's Folly, where Mayor Drummond presented them with a key to the town. One of the most pleasing parts of the day's program, however, was the gathering of former public school pupils in the Central

School at the sound of the bell at nine o'clock in the morning. Former principal R.J. Campbell took on the role of "teacher," and the "students" responded to roll call. The students spent a happy hour reminiscing about their school days. In the afternoon the cornerstone for a new high school was laid. A program of sports and races was also carried out with separate races for Windsor Old Boys and Old Girls. A street carnival and "old time concert" was organized for the evening, culminating with a "Night Shirt Parade" beginning at 11 p.m.

July 3 was designated "County Day," and July 4 "American Day" in honour of visiting Old Boys and Girls from the U.S., who were also celebrating American Independence Day. Parades, band concerts, and sporting events filled each day of the celebrations. On July 3 a girls' softball game was played between teams from Windsor and Petrolia in the afternoon. American Day, culminated with a spectacular fireworks display, which the program committee advertised as "a replica of fireworks started at the C.N.E. Exhibition." Street dancing and spontaneous street parades took place each evening. The people of Petrolia and their guests made the event their own, creating their own popular celebrations beyond those of the festival planners.

The final day of Old Home Week, July 5, was designated "Decoration Day." In the morning former ministers occupied the pulpits of the town's churches. A Decoration Service followed at Hillsdale Cemetery in the afternoon. The commemoration of the dead and the memory of those lost in World War I contributed to the rebuilding of community solidarity. In the evening an open-air service featuring a mass choir of 150 voices was staged at the children's playground. In the aftermath of World War I the commemoration of the dead and the building of war memorials to local sons lost in battle were replicated in small towns all over Canada. A particular modern form of memory evolved, built around the massive democratization of the cult of the dead, with the presence of saints and martyrs and a heritage to emulate.[25]

The 1925 Old Home Week celebrations, like the festivities in 1908, served to redefine public memory. In 1925 there was a need to reassert community identity in the wake of World War I and the social upheaval that followed. Unlike the 1908 celebrations, however, women were active in planning and staging the spectacle. The 1925 festivities were pronounced a success, and the *Advertiser-Topic* declared that "Old Home Week Will Go Down in History."[26]

Seventeen years later, at the close of World War II, a committee of Local No. 1 of the National Petroleum Workers Union employed at the Canadian Oil Refinery and the executive of the Central Lambton Exhibition Association called a public meeting to discuss plans for a third Old Boys' reunion. At the meeting a committee of both middle-class men and union members was constituted to organize the event. Another committee of Honorary Associate Members was formed, consisting of prominent middle-class men and women. The reunion was scheduled to coincide with the revival of the Petrolia Fall Fair after a lapse of several years. Unlike the previous reunions, organized labour and the farm community were given prominent public roles in planning the event. Any profits from the reunion were to be directed towards a fund designated for the construction of a Memorial and Recreational Park.[27]

The organizing committee sent a letter to all "The Petrolia Hard Oils" inviting them to attend the reunion scheduled for 26 August to 2 September 1946. Stories of the origins of "Hard Oil" circulate among Petrolia families to this day. "Hard Oil" is sometimes used to describe the petroleum itself, distinguishing "rock oil" or "hard oil" from other "softer" sources such as vegetable and animal matter. One widely told story attributes the expression "Hard Oil" to a rugby match in 1910, when a fan, J.J. Fisher, produced a fancy water pail for the team. The outside of the pail was trimmed in green and white, the Petrolia colours, with the players' names painted on it. On the inside of the pail were written the inspirational words "Hard Oil Finish." Another story attributes the origins of "Hard Oil" to the common community origins that bound together the Petrolia drillers who worked in foreign oil fields. In 1946, however, the "Hard Oil Spirit" symbolized the community spirit of Petrolians that pulled them together to overcome obstacles.[28]

The organizing committee issued a souvenir reunion program.[29] In addition to the schedule of events, the program contained two photographs of oil tanks, probably taken sometime in the late nineteenth-century. Both photographs are staged, one with a man and a woman posing on top of an oil tank in their best finery and the other of a lone man on top of an oil tank. The caption reads, "Do You Remember," with the following narrative:

> Many of the older people will recall these scenes of community life in the town when oil was flowing and storing was quite a problem. Refining processes were brought into being and the commercial production and use of oil was born from this very small beginning. To the younger people of today crude in its

structure, but it served its purpose and was the foundation of the big refineries and special processing which has come through years of research.

The program committee used the narrative of Petrolia's history as an innovator in petroleum technology to reconstruct public memory to embrace a younger generation, many of whom were probably employed in modern refineries and had only heard of Petrolia's origins as an oil producing community from grandparents.

Also included in the souvenir program was the historical narrative "My Petrolia," by Ann Blacklock, a local resident. Blacklock begins in the present with an announcement of the plans for the upcoming Old Boys' reunion scheduled to commence on 25 August 1946. "They are coming home. Home to Petrolia, to old friends, old scenes, old memories ... back to those intangible values which make a town live. Call it what you will, community spirit, good fellowship, loyalty, here we call it the Hard Oil Spirit. Where did we get the term? Roll back the years." Blacklock's history is a narrative of progress and triumph against adversity from first settlement by Europeans in 1847, through the oil boom of 1865, the incorporation of the town in 1874, and the invention of Henry Ford's "horseless buggy," which created a worldwide demand for the skills of Petrolia drillers. She recounts the first two Old Boys' reunions in 1908 and 1925. The Hard Oil Spirit waned somewhat during the Depression of the 1930s, when many families were forced to leave town. Nevertheless, Blacklock writes, "Petrolia suffered the depression in its own peculiar way," and turned the defunct railway depot into a Public Library. With World War II, the Canadian Oil Refinery expanded and the machine shops began working overtime. According to Blacklock, "the old Hard Oil Spirit revived," and "was fanned to a white heat by the zeal of National Petroleum Workers' Union, Local No. 1, of the Canadian Oil Refinery," and of Mayor Albert E. Thomlinson. Blacklock's narrative was more than a retelling of the story of Petrolia, it helped to rebuild public memory around the spirit of community that developed out of a shared history of struggles with adversity overcome through the "Hard Oil Spirit" of its citizens.

Following the example of the 1925 reunion, the *Advertiser-Topic* compiled a special Old Boys' Reunion edition intended to "bring a flood of memories to the old-timers of the community."[30] The editor drew on local geography and social relations built around an idea of place as home to structure remembrance:

Petrolia is lovingly remembered as the home town of most of our expected guests. Many will remember the old block pavement on Main Street, Stove Pipe Corner and the old swimming hole at Woodley's pond. They'll remember the mud on Tank Street and the slippery condition of the roads after a rain; they'll remember the sound of the school bell and they'll recall the names of their teachers with various degrees of warmth, and even affection, and forget how at one time they longed to "lick" the principal for some real or fancied wrong. Youth is that way. It is hot and it is forgiving. Many, too, will recall deeds of kindness by older members of the community, many now sleeping in Hillsdale and they will recall, too, the outstanding men of their day whom they thought of as giants only to learn, as they made their way around, that they were all about average.

Individual memory, as the editor's comments suggest, is subjective. At the same time collectively held ideas and experiences shared with others who spent their childhood in Petrolia formed the basis for a public memory of place as home.

Once again, Old Boys were asked to contribute their remembrances of their childhood in Petrolia. Len Howlett, who had been a resident of the town for more than 70 years, recalled his work as a teamster. At the age of 15 he acquired a horse and light dray and went to work for a local contractor. Five years later, he purchased another horse and set himself up as a teaming contractor, working as such for twenty years. He teamed for almost all of the local oil producers. Howlett's narrative focused on his experiences in the oil industry around the turn of the century.

Unlike the 1925 souvenir edition of the *Advertiser-Topic*, the recollections of an Old Girl were included in the 1946 reunion issue of the newspaper. Dr Mary Margaret Fraser, a former resident who had "made good" by becoming a doctor and organizing the first cancer clinic in Detroit, provided her memories of her childhood in Petrolia. In his introduction, the editor of the *Advertiser-Topic* wrote, "Citizens of this town are very proud of the former member of the local telephone staff, then popularly known as 'Queenie,' whose ambitions led her into the medical profession in which she has been an outstanding success." By locating Dr Fraser in Petrolia as young Queenie, the newspaper ordered public perceptions of the past and reinvigorated attachment to the imagined community.

Dr Fraser's remembrances were structured by her middle-class background and feminine gender identity, which focused on family, the home, and her affiliations with the women in the neighbourhood. She begins by attributing her professional success to her childhood, where the direction of her parents, the church, and her fine neighbours and friends moulded her thoughts and developed her ideals. She writes, "That cloak of protection has been given all who grew up in a small town, who know all virtues and felt the guidance of the people who developed and maintained a place like Petrolia." Fraser's reconstruction of the past was to some extent fashioned in the present around her professional success, but her self-identity was rooted in her lived past as a child in Petrolia. Thus the past is linked with the present.

Her earliest recollections were located in time and space in the neighbourhood around Crescent Park and Quality Hill, where the local elite lived in huge Victorian style homes. She listed the names of many of the prominent families of Petrolia who were her childhood neighbours. One of her earliest remembrances occurred when she was two or three years of age. She recalled venturing next door to Mrs Dibb's, where she found the house empty. She wrote:

> The table was set and awaiting the arrival of the wedding party of Miss Ruth Dibb, who was being married to Mr. Voe. When Miss Dibb arrived on the scene, she found me up at the table enjoying the wedding feast. My mother was chagrined but Miss Dibb wouldn't stand for any scolding because she said I knew I was welcome there at any time. You could not beat that for a child's friendship anywhere.

Mary Fraser makes fluid her childhood past and her adult insight thereby rendering the representation of the small-town-as-home visible to other adult members of the community, who probably shared many of these experiences. She continues in the same vein in the following paragraph where she describes the neighbours on the other side, the Moncreiffs, who encouraged her to practice the piano. "When I got tired," Fraser remembered, "Mrs. Moncreiff would make me a 'party' of bread and jelly and milk." Like the Old Boys who contributed their remembrances, Fraser also recalled her school days and ice skating in the winter.

The 1946 Old Home Week celebrations began unofficially at midnight on Sunday, August 25, when crowds gathered in the street, and formed a pyjama parade. Dancing and other forms of popular spontaneous celebrations emerged, suggesting that in Petrolia the hegemony of the festival's planners over public culture was challenged. The

Advertiser-Topic reported that "The crowd started to gather in town about 9 o'clock on Sunday night and gradually became more playful and a little noisier until a crescendo was reached about midnight with tooting horns, a series of explosions." Mayor Albert E. Tomlinson, his wife, and his children led the parade, all attired in night clothing. Informal pyjama parades, an "invented tradition" carried on from the 1925 reunion, were organized through the "grapevine" over the course of the weeklong event. Popular celebrations emerged during the 1946 reunion as the townspeople shaped the event to suit their own ideas of public spectacle. The *Advertiser-Topic* described these parades as "a combination of Hallowe'en, New Year's Eve, Mardi Gras and peace celebrations."[31] The reunion began the following morning with the registration of Old Boys and Old Girls. A parade of bands and a tattoo were held on opening night in Greenwood Park. Mayor Tomlinson officially opened the celebrations. Earl A. Smith, general superintendent of the Canadian Oil Refinery, addressed the crowd, pointing out the pleasure of renewing friendships. George Finlayson, president of the National Petroleum Workers Union, also addressed the gathering, stating that the Union's ambition was "to make a bigger and better Petrolia and to have Greenwood Park the best sport grounds in the region."[32] Unlike in the previous reunions, the labour movement had a visible official role in the 1946 reunion and played an active part in its organization. Local 1 of the National Petroleum Workers' Union also spearheaded the community drive to use the reunion as a way of generating funds for a Memorial and Recreational Park.

A Labour Day Parade was held on the final day of the reunion. Prizes were awarded for the best patriotic float, the best industrial commercial float, and the best comic float. The main attraction at the Labour Day celebration was the shooting of an oil well on the grounds of Greenwood Driving Park. A committee chaired by Dr Charles Oliver Fairbank, the son of pioneer oil producer John H. Fairbank, organized the event. The well, which was christened "The Morning Glory," was drilled prior to the reunion under the direction of veteran driller, William J. Hussey, with the assistance of several foreign drillers, namely, Fred H. Edward, William Cole, Bloss Sutherland, and Ernest Kells. As in the previous reunions in 1908 and 1925, the role of the foreign drillers as global experts in oil drilling and in shaping Petrolia's identity around the oil industry was showcased.[33] Local businesses, which were "pioneers" in the oil industry, donated equipment for the drilling operations: William Hussey supplied the drilling rig, the Oil Well Supply Company supplied drilling tools, VanTuyl & Fairbank provided the casing, and J. & J. Kerr Company supplied the lumber. The Canadian Oil Company, Imperial Oil, and British-American Oil supplied the oil and gasoline for the operation of the drilling rig.[34] Oil was struck a few days earlier, on September 3, at 469 feet. One hundred pounds of dynamite were lowered into the hole in three large metal containers. At four o'clock in the afternoon the charge was set off, and a gusher of oil sprayed 100 feet into the air to the delight of the estimated 10,000 spectators.[35]

A midway and a sports program were incorporated into the 1946 Home Week celebration. Men's baseball and women's softball games were held every day. A refreshment tent was set up in Greenwood Park where meals were served by the various women's groups in the community, including the ladies' auxiliaries of the churches, the women's institutes, and the Ladies' Auxiliary to the Canadian Legion. The Ration Board provided extra rations for these meals.[36] The Petrolia and Enniskillen Fall Fair, which had been abandoned in 1928, was reorganized in conjunction with the 1946 Old Home Week. The celebration incorporated the rural farm community, both as organizers and participants. Sunday, September 1, was designated Memorial Day, to commemorate the founders of the town and those who had given their lives in the two World Wars.

One of the most popular events with the Old Boys and the Old Girls was the "holding of school," another vestige from the 1925 reunion. The *Advertiser-Topic* commented that "Memory doesn't forsake the times and people of school days." The teachers, however, were remembered differently. Where they might have been remembered as stern taskmasters and disciplinarians, "the teachers of these same pupils are held in a new reverence and wear imaginary halos in the minds of the former pupils."[37] Two sessions of school were held at the Central Public School. The classes were conducted by former teachers.

Following the 1946 reunion the editor of the *Advertiser-Topic* wrote, "The old home town has a mysterious and powerful appeal such as no other place can match or begin to equal. While people grow older, new faces appear and buildings change, the sentiment attached to the community remains as memory keeps green the happy associations of former days." The analysis of the three Old Boys' reunions suggests that place was essential to the process of constructing public memory. The identification of Old Boys and Old Girls with place-as-home was part of a social process whereby individual childhood memories formed the basis for building public memory. A shared past was built around the local history of Petrolia as an oil resource town. The oil derricks that covered the countryside became community symbols. The Old Boys' celebrations also illuminate the political nature of history and memory. The 1908 reunion was organized by the middle-class men of Petrolia for their own commercial gain, suggesting continuity with the civic boosterism of the nineteenth-century. The celebrations also displayed the public in gendered imagery. The event was organized by men, conducted by men, and depicted using masculine signs. Women were visible as organizers of the 1925 and 1946 reunions, although they were given the "municipal housekeeping" roles usually assigned to women. The 1946 reunion marked a departure of the domination of the event by Petrolia's middle-class male businessmen with the active involvement of Local No.1 of the National Petroleum Workers' Union and the rural farm community in organizing the event. In all three Old Boys' reunions, spontaneous public revelry emerged, suggesting that the hegemony of the predominantly middle-class elite in planning and carrying out the Old Boys' reunions was always contested.

THE PETROLIA DISCOVERY: CANADA'S HISTORIC OIL FIELD

By the late 1960s Petrolia's infrastructure, much of it built during the oil boom days of the nineteenth century, was badly worn and poorly maintained. The community lost its remaining refinery when Canadian Oil transferred its operations to a new plant at Froomefield on the St Clair River in April 1952. The deep wells on the old Petrolia site were turned into a waste dump for caustic acids from Sarnia's Chemical Valley. Confronted with the prospect of creating a new competitive nucleus of local production and consumption, heritage in Petrolia shifted from an emphasis on public spectacle and the creation of model citizens, as evidenced by the three Old Boys' reunions, to a focus on tourism and the model consumer. The phenomenon of place promotion was not new; it was at the forefront of each of the three Old Boys' reunions; but beginning in the 1970s it was without a doubt more market driven.

In Petrolia, as was often the case in towns and cities throughout the Western world that confronted deindustrialization in the 1970s, much of the renewed interest in heritage was initiated at the local level. It emerged in a particular local context that involved public agencies and community organizations with a variety of motivations, not all of which involved free enterprise.[38] A Promotional Committee of the town council was organized in January 1972. Robert MacFarlane, a member of the committee, proposed the establishment of a demonstrational oil field and picnic ground in Petrolia to supplement the Oil Museum of Canada in Oil Springs. In February 1973 the town council picked up the option on the property owned by Murray Bradley, a 50-acre parcel of land east of Tank Street and south of the Blind Line; the site of the original Petrolia oil discovery in the 1860s. Bradley, who owned and operated the oil field from 1949 to 1972, sold the property to the town for $10,000. The remains of the Fitzgerald Rig were also on the property, which added to the attraction of the site for a living museum. The Fitzgerald Rig was named for Frederick Ardiel Fitzgerald, the first president of Imperial Oil, who constructed the building in 1903 to provide a cheap source of power for the Petrolia field. The original steam-powered engine and its two gigantic bull wheels operated as many as 300 rigs by jerker rod. The committee urged that the property be developed soon because old equipment was being lost and only a few of the "old time" oil producers were left with first-hand knowledge of the system.[39]

The historic oil field, the conservation of the downtown Victorian façade, and the restoration of Victoria Hall as a venue for summer theatre were viewed by the members of the town council as potential tourist attractions that would create a new competitive nucleus of local production and consumption. Also at stake was the survival of place identity built around a local history as Canada's Victorian oil town. The Promotion Committee prepared a preliminary budget of $20,000 for the development of the historic oil property, and began searching for funds. In July 1973 they asked both the county and the town councils to finance the project. The *Advertiser-Topic* hinted at some dissension between the committee and a group of community volunteers, Victoria Playhouse Petrolia, which was seeking funds to restore Victoria Hall. Members of both groups denied that there was any conflict of interest between the groups, although one unnamed member of the playhouse committee suggested that both groups should do more to inform each other of their fund-raising activities.[40] A separate Historical Oil Site Committee was established to avoid any further allegations of competing interests.

Figure 6.2 The Petrolia Discovery. Personal collection

In December 1973 a grant of $15,320 was awarded under the joint federal-provincial Local Initiatives Program to develop the oil-well site. Local initiatives programs and other forms of public funding functioned according to a competitive bidding process, which helped to transform the role of local government into one of place promotion by generating images both of a traditional industrial past as well as a dynamic post-industrial renewal. The funds were used to clear the site, move equipment and buildings onto the site, and begin the restoration of the Fitzgerald Rig.[41] The Historic Oil Site Committee, chaired by Gary Ingram, canvassed key oil refiners, including Imperial Oil and Shell Oil, with refineries in Sarnia, for funds. Lambton MPP Lorne Henderson and federal MP Robert Holmes were unsuccessful in their efforts to secure a grant from the National Museum of Canada fund established to promote branches of larger museums in small communities. The historic oil field project languished for lack of financial resources.[42]

The *Advertiser-Topic* lamented that soon there would be no relics to display anyway. Popular nostalgia for the past brought antique collectors from all over Canada and the U.S. to the oil field. "When the writer came here 14 years ago, the fields around Petrolia and Oil Springs were literally dotted with tripod derricks, the symbol of the early oil industry. Today there are only five or six left," the reporter remarked. Drill poles were disappearing and antique collectors had stripped the woods of remaining jerker rods. Only three or four men were left who had any knowledge of early drilling technology.[43] Preservation initiatives by the Historic Oil Site Committee were presented as a legacy at risk and were bound up with the feeling of a loss of a way of life. Heritage enthusiasts were also motivated by a concern for the survival of place distinction.

Brian Arnott, who was a consultant on the restoration of Victoria Hall, was hired to supervise exhibit development for the historical oil site project. The original concept of restoring the site as a "pioneer oil field" expanded to include an "industrial pioneer village," recreating the living and working environment of those who established the oil industry in Petrolia. The industrial pioneer village was intended not only to preserve the technology of the oil field but also to allow tourists to experience living in a pioneer resource town by walking through reconstructed streets and conversing with real costumed performers. Arnott's recommendations were consistent with a broader trend beginning in the late 1970s and 1980s, where cultural display is redirected at a mass audience and "living history" sites display a vernacular past. These heritage displays provide live representations for tourists, but they also allow local people to ensure that their own traditions continue, ones that otherwise would probably be lost.[44]

A grant from the Wintario Lottery Corporation, and a vigorous fundraising campaign by the members of the Historic Oil Site Committee were pivotal to the opening of the Petrolia Discovery. Funds to restore oil wells were raised by "selling" them at $15,000 each; corporations or private donors could have a well restored in their name for a cash donation. By July 1979 four wells had been "purchased," two by a petroleum corporation and two by private foundations.[45] Over the course of the summer, three wells were restored at the site. Local oilmen donated their time to ready the wells. At first the wells produced only water, but by early September they were generating about one barrel of oil per day, with crude selling at $14.31 per barrel. The revenue generated from the wells was added to the restoration fund and later helped with operation costs after the museum officially opened, thus providing the museum with an income independent of tourist and visitor revenue and public funding.[46] In September 1979 Imperial Oil Limited donated $70,000 to the project to commemorate the centennial of the relocation of its refinery in Petrolia in 1880.

In the summer of 1979 a crew from the National Film Board began work on a documentary about the early oil industry intended for visitor viewing at the Petrolia Discovery museum. The shooting of live footage by cameraman-director Nick Kendall took place around Oil Springs and at the Discovery site in Petrolia over a five-day period in December 1979. In addition to the film footage, the filmmakers used a composite of still photographs and postcards collected from area residents. Produced by David Springbett, "Hard Oil" had its premiere at Victoria Hall on 27 June 1980. It was attended by approximately 600 people, many of whom had ancestors who worked in the Enniskillen oil fields. Following the screening Mayor Crawford enthusiastically pronounced that "'Hard Oil' could be Petrolia's 'Sound of Music'."[47]

The text of "Hard Oil" draws on meta-narratives that are easily recognized and readily comprehended by the tourist gaze. The particularities of local history are filtered through mythic narratives of community-building and everyday experience. The story of the oil industry is presented as a story of modernization and industrialization. The nature of the local oil industry is described through the familiar trope of hard work resulting in wealth and an easier life for the next generation. "Despite its rough edges Petrolia grew into a solid Victorian town." "The founders were black with oil; but their children had time for other pursuits such as needlepoint, theatricals, sports and more sports."

Petrolia's national and global frame is reiterated throughout the film. With nationalistic pride it is claimed that James Miller Williams brought in the first oil well in 1858, before the Drake well was struck in Titusville, Pennsylvania, in 1859. Petrolia is represented as the "little town that kept the whole country running." Later in the film, a male narrator depicting an Old Timer declares: "Forget Alberta, Saudi Arabia; we had the first oil well ... of course we were proud." The technical know-how of the foreign drillers is highlighted and further establishes Petrolia's connections to the rest of the world. The sense of a unique place-identity is heightened also: "Petrolia invented the oil industry – we were really something." Thus, Petrolia is transformed into the epicentre of wider social and economic forces. Without Petrolia and the technical expertise of its oil drillers there would not be a global oil industry. The text of "Hard Oil" works to connect the local story of Petrolia's oil boom to world history, where Petrolians are given their long-overdue recognition and a unique place-identity is heightened rather than dissolved in the overarching sweep of globalization.

The Petrolia Discovery Foundation was organized in 1980. Hugh Lee, a twenty-four-year-old geography graduate from the University of Western Ontario, was hired as site manager. Brian Harrison, who had already been at work for several months, was employed as a blacksmith, and Frank Belford operated the oil field. In March 1980 the Petrolia Discovery received its first instalment from Wintario and a full-time bookkeeper was hired. Local fundraising projects plus the Wintario grant allowed the Project Discovery Committee to open up tenders for the construction of an admission building and sales outlet, six open display areas, boardwalks, a raised platform so visitors could get an overall view of the site, and a movie pavilion, where "Hard Oil" would be screened for visitors. Local Rotarians donated their time to help with painting on the site, and the Petrolia "Krazies," a community group of young adults, planted trees.[48]

A Technical Committee of twelve local men with expertise in the oil industry was established. Albert Baines, the owner of the Oil Well Supply Company, chaired the committee. The committee was responsible for the actual restoration of the historic oil field; they were responsible for cleaning and pulling wells, erecting three pole derricks and setting-up and maintaining jerker lines and wells. The restoration of the oil wells involved modern interventions. As historian Raphael Samuel writes, "living history is a matter of improving on the original by installing, in replica and facsimile, what ought to be there but is not."[49] Objects must be restored to some life-like replica of their original habit if they are to be intelligible in the present. In May 1980 the *Advertiser-Topic* reported that "Authenticity is important at Discovery," but that it was "not the easiest chore." In the end the Committee decided that the restoration would be "authentic-looking." The poles for the derricks were constructed from "ancient" telephone poles from the town of Petrolia, instead of black ash trees cut from the swamps of Enniskillen Township, which were used in the original construction of derricks in the nineteenth-century. The wood on the site's jerker line system could not be patched and a decision was made to replace the system with a fascimile. Also, wooden pump jacks were used during the pinnacle of Petrolia's oil days, but only a few wooden pump jacks were used in the restoration. "It's a compromise," site manager Hugh Lee remarked. "The Place has to be functional as well."[50] The objective of the Technical Committee was to make the past comprehensible in the present by rescuing relics decaying in local oil fields and restoring them to their previous function. Thus, it is not the original that appears "authentic," but current views of what the past ought to look like to make it comprehensible to the modern tourist. The past is staged for a tourist gaze that seeks out spectacles of ordinary life. As Bella Dicks explains, "They stage the vanishing of the familiar, but only through the familiar having already become strange."[51]

On 25 June 1980, three days before the museum officially opened, an editorial appeared in the *Advertiser-Topic* entitled, "Discovering Petrolia, discovering today." In addition to the tourist gaze, the living history project at the Petrolia Discovery invited the self-gaze of those on exhibit by ensuring that their own unique traditions continue. The editor draws on the narrative of Petrolia's pioneering role in the development of the global oil industry to highlight the value of their own place-based culture; a culture he believed was at risk of disappearing. The editor writes, "It's suspected that a great many Petrolia residents have yet to take a short drive out to the Blind Line at Tank Street and discover the town's heritage – and the world's – as seen in the Petrolia Discovery." The birth of the petro-chemical industry, he continues, was on the site of the Petrolia Discovery. "We were the Western hemisphere's Kuwait long before the Middle East was anything but sand and camels."[52] In Petrolia, globalization fostered cultural particularism and an awareness of the importance of locality.

At the opening of the Petrolia Discovery on 28 June 1980, a simulated "oil gusher" was set off at the entrance under a tripod derrick. The "oil," actually coloured water, was shot upward from a system of pipes devised by members of the Technical Committee. Visitors to the site were able to form a perception of the past through the sights, sounds, and smell of a working oil field. Where heritage is used for tourism, emphasis is placed on interactive experience rather than on simply viewing an object. The belt and pulley driven machinery driving the restored Fitzgerald Rig was the main attraction. A fully operational blacksmith shop was also installed and originally housed in the same building as the Fitzgerald Rig. By opening day seven wells had

Figure 6.3 Panorama of the Petrolia Discovery site: A Living Museum. Personal collection

Figure 6.4 The Fitzgerald rig. Personal collection

been restored on the property. Seven students were hired as guides under a federal government Young Canada Works grant. To prepare themselves for the job they took a two-week immersion course in local history. All of the student guides were from the area, but many of them reported never having heard of the Fitzgerald Rig and few remembered ever having visited the oil field.[53]

The historic oil field was conceived as an interactive display intended to explain the pioneer oil technology developed in the Petrolia oil field. As manager Hugh Lee explained, "Petrolia Discovery will be developed with a balance between the educational and the fun aspects of a tourist attraction. People will be able to learn by seeing and smelling a real oil well in operation. It will be a total experience."[54] What is advertised as "Canada's historic oil field" is presented as a narrative of progress in oil drilling and refining technology that embellishes Petrolia's claim to national and international prestige as pioneers in modern oil drilling technology. Visitors pass a facsimile of the gum beds found on lots 17, concession two, of Enniskillen Township by Charles Tripp in the early 1850s. In actuality, the gum beds were located on the site of the first oil boom at Oil Springs. A large iron kettle, similar to the one used by early oil refiners in Enniskillen Township, is situated next to the gum bed exhibit. Visitors then pass a dug well, also called a cribbed well, before moving on to the spring pole exhibit, where they can jump up and down on the spring pole much as the pioneer oil men did in Enniskillen Township during the early 1860s. The Canada Rig showcases the drilling technology that Petrolia drillers took to foreign fields beginning in 1874. The Canada Rig was constructed on four poles and was easily mounted on skids and drawn by horses to drilling sites. Alongside the rig is an exhibit of drilling and fishing tools, most of them manufactured locally at the Oil Well Supply Company. Visitors can also experience a sanitized version of a pumping well. The Fairbank Well, named after pioneer oilman and "Father of Petrolia," John H. Fairbank, is pumped by a wooden jerker line powered by the Fitzgerald Rig. The historic oil field is pristine, unlike the actual grease-covered oil fields of the nineteenth-century, in keeping with modern tourists' expectations that public spaces be consumer-friendly, accessible, interactive, and safe.

A series of pavilions representing the wooden shanties found in Petrolia in the 1860s line a wooden boardwalk. The first panel is devoted to the Imperial Oil Refinery, which is not surprising given the corporation's large financial contribution to the museum project. The exhibit highlights Imperial's origins in Petrolia, but manages to collapse the past into the present while at the same time suggesting more capitalistic intentions, with the caption "Sharing a Past, Investing in a Future." No mention is made of the huge social and economic impact the relocation of Imperial Oil to Sarnia in 1898 had on the community. One pavilion contains photographs of Petrolia during the first decade of the oil boom from 1860 to 1870. The exhibit coordinators selected photographs of the town's main street, lined with bars, barbershops, and wooden hotels, which depict Petrolia during its early days as "a man's town." Another pavilion named "Mrs. Cole's Kitchen," is a replica of an early twentieth-century kitchen, thus highlighting women's domestic work in the home. "The Foreign Driller" pavilion describes the Petrolia drillers who brought local drilling technology to oil-rich areas in more than fifty countries. A world map illuminates the extent of Petrolia drillers' travels. The NFB film "Hard Oil" is played for visitors in the sixty-seat theatre constructed on the site. The museum is organized around a nationalistic narrative of Petrolia as "Canada's pioneer oil field," and a linking of the local and the global around the theme that Petrolia drillers played a pivotal role in the discovery of most of the major oil fields of the world. An advertising pamphlet states, "Before there was OPEC, lipstick and nylons, Iran and Alberta, there was Petrolia. Here, the world's oil industry started, and it's waiting for you."

Since the early 1980s the re-creation of a pioneer industrial village has been an ongoing project at the Petrolia Discovery. The idea of "living history" involves the creation of an imaginary past. At the Petrolia Discovery an imaginary past was created from a nucleus of material already on the site by assembling artifacts on the site and threading them together using the theme of the "pioneer industrial village."

In the fall of 1980 Imperial Oil Limited donated 300 blacksmith and oil-related tools to the museum. The following year the Imperial Oil Pumphouse, part of a shipping and receiving station, was moved to the Discovery site from Bunyan in neighbouring Sarnia Township. The pump inside the building and the gas engine powering the pump were included. The pump was used to push oil from the Petrolia oil fields to Sarnia during the early decades of the twentieth century.[55] In July 1982 an 112-year-old house designated for demolition was moved from Tank Street to the Discovery site and transformed into a blacksmith shop, with the aid of funding from the local chapter of the IODE and a grant from the Lambton County Council. The aim of the resto-

Figure 6.5 The blacksmith's shop, Petrolia Discovery

ration project was to recreate an "authentic" replication of a nineteenth-century oil field smithy, and the structural configuration of the house, which was originally constructed using the vertical batten method, was deemed suitable. During the course of the renovations, a brick chimney and base and a replica of a Victorian-era forge were constructed. In developing the concept of a "pioneer industrial village," intended to replicate Petrolia during its boom-town days, an old building acquired a past at odds with historical reality. Remnants of decorative wallpaper remain inside the reconstituted blacksmith shop. Nevertheless, the recreation of the nineteenth-century blacksmith shop was conceived as a working shop in keeping with the concept of a living museum. A blacksmith works at the forge making holders for the jerker lines, linkages for the pump jacks and souvenirs for the tourists. He explains the trade to tourists and schoolchildren. The blacksmith's speech and know-how are mostly modern, however. Brian Harrison, the first blacksmith on the museum site, learned the trade at a community college course in Alberta. On the job he was equipped with a modern plastic safety helmet.[56]

As the idea of a pioneer industrial village unfolded at the Petrolia Discovery during the 1980s and 1990s it took on the characteristics of "a modern collage" of buildings from different eras. The buildings were removed from other sites as a mode of historical salvage and were arranged as though they existed door-to-door on the same street. The buildings were all restored using vertical boards and batten, and they were painted black with red trim on the window frames; a recognizable symbol for the tourist gaze of the rustic, pioneer industrial town. According to Bella Dicks, the visual effect of what she describes as "theming culture," is to recreate a simulacrum of reality by deploying a recognizable set of symbols and eliminating the "dead space," or the non-symbolic, ordinary, unreadable spaces which real environments always contain.[57] "Theming" also suggests a visual uniformity far removed from the historical era the museum's organizers sought to replicate.

In 1990 a railroad shed used as a warehouse by the Canada Central Railway, and dating back to the 1870s, was moved onto the site. The building was converted into a wood-working shop, "to preserve and illustrate the techniques used to fabricate the wooden materials of the oil industry at the turn of the century."[58] The building, which was opened to the public in May 1993, was re-named Watson's Mill. Most of the woodworking equipment inside was purchased in 1989 at an auction in Watford when Russell Watson went out of the lumber business, hence the new name. A distribution building, dating from the 1920s, and later purchased by Harrison Corey in 1942 when Corey Oil was established, was moved onto the Discovery site in 1992. This building too was refinished in board and batten in keeping with the idea of recreating a facsimile of a pioneer main street. Photographs of Petrolia during the oil boom era of the nineteenth century are displayed in the Corey Building to provide context for the pioneer industrial village. The building is also used for crafts when groups of school-children visit the museum.

In 1994 a one-room schoolhouse was donated to the Petrolia Discovery. The schoolhouse, built in the 1870s, was originally on lot 12 east, concession nine, in Enniskillen Township, and was moved in the early 1900s to lot 16, concession nine. The original flooring, tin ceiling, light fixtures, and slate chalkboards were restored to approximate a turn-of-the-century schoolhouse. A deck and wheelchair accessible ramp was added to meet the needs of visitors. Former students helped in the renovation of the schoolhouse, which was fitted with early desks, books, and other related artifacts. A reunion of former teachers and students was held on 14 September 1995 to celebrate the opening of the schoolhouse exhibit.[59] The schoolhouse also serves as a historical teaching facility, with actors re-enacting the role of "school marm" in period costume. A schoolhouse program, the result of two years of planning by a committee formed by the Lambton County Principals' Association, was introduced at the museum in 1998. The two-hour program is conducted as a re-creation of an elementary school class around 1900 "complete with the religious aspects which are so controversial today." Discovery site manager Betty Popelier remarked, "There is religious content in the program, because that is the way it was done then." While the actors no doubt fill in the gaps with the probable and the invented, the re-enactments enliven history for schoolchildren, who find the teacher "kind of strict."[60]

In July 1998 a one-storey frame house thought to be one of several toll-road houses, circa 1880, was moved from Sarnia to the Discovery site. The building was renovated as "The Petrolia Heritage House." It houses artifacts found in a closet at Victoria Hall and any other relics found in the attics of local homes that deal with the social history of the town and "reflect the opulence that was characteristic during the oil boom days of the late 1880's."[61] In the fall of 2000 the Methodist Episcopal Church, built in the 1880s on the Marthaville sideroad, was moved to the Discovery site.

The Petrolia Discovery, as a living museum, has had to conform to modern environmental regulations, and thus departs from the reality of oil production in the nineteenth century. Ninety-nine per cent of the production from the Discovery field was brine. The traditional nineteenth-century practice was to dump the waste material into the watercourses. In the summer of 1988 the Ministry of the Environment threatened to close down the historic oil site unless an injection well was constructed. A brine disposal well was constructed at a cost of $60,000.[62] The brine flows through a pipeline into a disposal well, where it is fed into the porous formation beneath the oil field.

The redevelopment of the Petrolia oil field was interwoven with state environmental protection initiatives and the conservation movement that spread across North America beginning in the 1970s. In 1982 the CRUDE program – Children Raising and Understanding Deer in their Environment – was initiated. Children from four local schools were given the responsibility of raising and caring for a small herd of deer. In the early days of the oil industry, deer were a common sight on local oil properties. The intent of the program's organizers was to allow children to learn about deer in their natural habitat. A rancher from Blenheim donated four Sika deer. Unfortunately, three of the deer escaped through holes in the enclosure. The sole remaining deer was named Lonesome.[63] A goat and two llamas were placed in the pen with Lonesome. Eventually, the Discovery Board decided that the connection between livestock and petroleum was too "tenuous" and the project was abandoned.

At the Petrolia Discovery nature was put on display with the addition of the William T. Abraham Arboretum in 1994. Before his retirement in 1983 William Abraham served for 25 years as Lambton County's representative to the Ministry of Agriculture. The members of the Discovery Foundation were eager to preserve as many of the trees that are unique to the area as possible. Many species of trees disappeared in Enniskillen Township beginning in the mid-nineteenth century when land was cleared of timber to establish farms and poles were required for oil derricks. Volunteers have planted more than 105 trees of at least a dozen species that are indigenous to the area, including bladdernut, black walnut, silver maples, and several species of ash and hickory.[64]

In organizing heritage as tourism, the Petrolia Discovery has taken some of the lessons from theme parks and has designed exhibits that are interactive, thus enabling visitors and tourists to experience a facsimile of life in a pioneer industrial village. Special events are also held on the site, including the annual antique car rally, a pumpkin festival, and Victorian Christmas. The restoration of the historic oil field and the creation of a pioneer industrial town involved a combination of public funding, local volunteers, professional consultants, and local oilmen with technical expertise. Each of these groups had its own ideas about how to stage heritage for tourism. Local men and women had considerable input into how their history was exhibited and how traditions that otherwise might be lost were maintained. Local involvement was strong, particularly where local history was used for educational purposes. The Petrolia Discovery is a unique museum in that it has a source of revenue aside from admission fees, government grants, and donations from the public. It is a working oil field and each month the crude produced is sold to Imperial Oil, and the income is used to pay operating expenses for the museum. Nevertheless, during the early 1980s the Petrolia Discovery operated at a deficit. Bingo games were held to reduce the deficit, and the Town of Petrolia and the Toronto-Dominion Bank forgave loans, so that by 1985 the museum was financially solvent. In 1990 a unique annual fundraiser, called the "Black Gold Raffle" was introduced. The winner is entitled to a ten-year income of one of the site's oil wells, and selects a name for the well. The chance to become an "oil baron" or "oil baroness," has been popular with the public.[65]

It is now possible to take a virtual tour of the Petrolia Discovery, with a map and photographic representations simulating each site in the historic oil field and each panel on the boardwalk. While internet tourism has the potential to create virtual destinations capable of being the tourist experience itself, the Petrolia Discovery seeks to use the website to attract visitors to the museum. The website contains a special events page, a contact page with admission fees and operating dates, and links to Victoria Playhouse Petrolia and other destinations around Lambton County. Place promotion and appetite whetting is the objective: to persuade people to leave their computer and visit the Petrolia Discovery in person.[66]

THE RESTORATION OF VICTORIA HALL AND THE CREATION OF "CANADA'S VICTORIAN OIL TOWN"

Completed in January 1889, Victoria Hall was built in the centre of Petrolia's main street. Victoria Hall housed the municipal offices, town council chambers, fire department, division court, town clerk, chief of police, town engineer, and an elaborate opera house that symbolized the town's prosperity based on the oil industry. The building was designed by George Durand, a well-connected and highly respected architect from London, Ontario. In addition to Victoria Hall, Durand was responsible for the architecture for the town's Presbyterian and Catholic churches and the Masonic Hall. During his ten-year career from 1879 to 1889, Durand's firm designed a number of buildings throughout southwestern Ontario, including the mansion of John Labatt in London and the second building of Upper Canada College in Toronto. Durand's design for Victoria Hall was a mixture of Queen Anne architecture and the vernacular. He responded to the demands of Petrolia's "oil barons" for a building that would showcase the town's wealth with a towered building in white brick, with entrances in set-back sections, each with distinctive facades, and fancy joinery. Above the wooden gazebo containing the fire alarm bell is a bell-case roof with four dormers sheltering the faces of the town clock.[67]

During the 1890s the fancy dress balls organized by the town's elite were held in Victoria Hall. Travelling professional theatre troupes

Figure 6.6 Victoria Hall, c. 1900s. Personal collection

staged performances in the hall.[68] The first professional performance was "Dr Jekyll and Mr Hyde," staged shortly after the building was completed in January 1889. Guy Bros Minstrel Company visited Petrolia on several occasions during the 1890s. Performance artist, poet, and native rights' advocate E. Pauline Johnson also performed in Victoria Hall. By the 1920s the heyday of Victoria Hall was at an end as other commercial amusement vied for the attention of the town's residents. Over the next several decades the hall was used only sporadically for local theatrical performances and Saturday night dances, and its splendor began to fade.

In the fall of 1972 a committee of prominent local citizens headed by Ron Baker, the librarian for Lambton County and an amateur actor and director, was formed to restore the Victoria Town Hall and Opera House as an operating theatre. The committee hired Toronto theatre consultant Brian Arnott to survey the premises. Arnott reported that it would cost as much as $100,000 to put Victoria Hall into a condition that would accommodate a professional theatre. Such a project, he noted, would require the support of the entire community. To bring the opera house up to modern theatre standards the roof of the stage would have to be completely rebuilt to take the heavy loads imposed by modern lighting and staging equipment, and a control booth would have to be added. To meet fire regulations a fire exit would have to be constructed from the stage, and more washrooms were needed to comply with health regulations. Arnott indicated that the opera house was well worth renovating. "There is no reason why Petrolia should not repeat the success of the Shaw Theatre at Niagara-on-the Lake," he stated. "This is a charming town with a rich history and the opera house should seek to draw out the historical values." He further recommended that the restoration of the opera house as a culture and community centre be carried out alongside a larger plan to restore the town core.[69]

A strong sense of place also fuelled the restoration project. "There is a very great need to give Petrolia a new reason for being, a new confidence in their town," the *Advertiser-Topic* stated. "In these days of centralized governments and centralized industry, we are becoming too much a bedroom community." Archeologist Kevin Walsh argues that the twentieth-century trend towards globalization has led to a need to protect that which stood as a metaphor for the characteristics of the nation. The preservation of Petrolia's nineteenth-century buildings was tied to the growth of Canada as an independent nation. As the *Advertiser-Topic* remarked, "It has always been our contention that it was here in Petrolia that the 20th century began for Canada."[70]

The restoration of Victoria Hall was part of a wider trend beginning in the 1970s that brought about the use of heritage for tourism. It was also suggested that the restoration of Victoria Hall might also serve the public need for a community centre. In March 1973 Victoria Playhouse Petrolia (VPP), a non-profit foundation, was established to raise funds for "the restoration of Victoria Hall to its former dignity and the return of the building to its original use as a community centre for Petrolia and surrounding Lambton County." As a charitable foundation VPP was eligible to apply for federal and provincial funding. A few months after it was organized, the VPP leased Victoria Hall from the town for five years for the nominal fee of one dollar per annum.

The Promotion Committee of the town council invited the Advisory Committee of the Architectural Conservancy of Ontario, a Toronto-based organization of citizens and architects interested in preserving historical buildings in Ontario, to conduct a survey of historical buildings in Petrolia. Peter Stokes (who was involved in the restoration of Niagara-on-the-Lake and Upper Canada Village), Toronto architects Howard Chapman and Howard Walker, and Doug Johnson of Windsor visited Petrolia in May 1973. In their report to the town council the following month, the architects remarked that "Petrolia presents a varied and fascinating collection of Victoriana, with the late Victorian period well represented." They further noted that "Good examples of the Victorian detailing referred to previously can be enjoyed in the exterior treatment of the Victoria Hall, the Masonic Hall, the Fairbank House and St. Andrews Church." The architects recommended that the main street, Petrolia Street, be preserved in its entirety to maintain the overall unity of the street. "A total scheme would revitalize the main commercial core and increase its attraction for the citizens of Petrolia and visitors," they wrote. This program of downtown revitalization would soon be tied to the development of the living museum at the Discovery site.[71] As the *Sarnia Observer* commented, "What started as an idea to put a fire escape and washrooms in the opera house on top of the town hall has grown into the planned festival and a whole new plan for the community.[72] The entire town was envisioned as a tourist attraction and the preservation of Petrolia as a historical town assumed the proportions of a major industry.

The town council voted unanimously to endorse the restoration of the Victoria Opera House and applied for a winter works grant to cover labour costs for the next two years. In July 1973 Petrolia town council received word that a $138,500 federal-provincial forgivable loan had been awarded. The grant, however, had restrictions; it applied only to work done during the winter months from 1 December 1973 to 31 May 1974. Only 50 per cent of the loan was forgivable for work done in the summer and fall. The balance of the money needed to restore the opera house, some $90,250, was raised by the VPP foundation through public donations.[73] Toronto architect Howard Walker was hired and by late October 1973 the architectural and theatrical drawings were complete. In 1975 Victoria Hall was designated a national historic site and monument and a heritage site under the Ontario Heritage Act.

Within the community tensions mounted between those who supported "culture" and the restoration of the opera house as a building of historic interest and a performing arts centre and those who advocated the use of public funding for sports facilities. One anonymous man, a volunteer in local amateur hockey, remarked, "Who wants to go to an opera anyway. I certainly wouldn't and I don't think most people would." For this man live theatre was boring; he would rather attend a Junior B hockey game at the local arena. The local arena still had not been completed nearly ten years after construction began, and the building was in need of a new floor, additional heating and routine maintenance. He pronounced the expenditure of public funds for the restoration of Victoria Hall a waste of money. The *Advertiser-Topic* countered "The money we spend for cultural development, for broadening our minds is not wasted." Money had little to do with it. According to the *Advertiser-Topic*, the goals of the VPP were much higher, namely the broadening of the minds of the citizenry so that the problems of social change might be understood and communicated. In the weeks that followed the *Advertiser-Topic* emphasized that the "Opera house project is really a community centre." The restoration of the town hall was pitched to the public not simply as a promotion of live opera theatre and a tourist attraction, but as the "provision of a community centre of a size and kind almost without parallel in this part of Western Ontario."[74]

The purview of the VPP soon extended beyond the restoration of a historic building to incorporate the presentation of special events in the opera house, as heritage in Petrolia became more closely aligned with market forces. Paul Thompson's Theatre Passe Muraille's production of "Them Donnellys" appeared in the opera house on 19 and 20 November 1973. The story of the brutal murder of the Donnellys, an Irish-Catholic family who settled in Biddulph Township near the Irish-Protestant community of Lucan, was well known in Petrolia, having been made popular by Point Edward clergyman The Rev. Orlo Miller's 1962 historical novel *The Donnellys must die.*[75] In just nine days, the volunteers from the VPP sold 1,390 tickets, and the shows were sold out two weeks prior to the performances. Seating was restricted to 460 seats per show in accordance with existing fire regulations. In his review for the *Advertiser-Topic*, Charles Whipp remarked that the "Donnellys is part of a new Canadian theatre." He praised the troupe's director, Paul Thompson, for his efforts to establish a totally Canadian theatre – one that is entirely democratic. Every actor contributed to the play; it was composed as rehearsals progressed without a script. Whipp praised the VPP for bringing this new kind of theatre to the opera house. "We might have been soothed with "Charlie's Aunt" or some other safe old chestnut to woo support for the restoration of this moldering hall," Whipp wrote. "Instead, VPP dared to present a naked, fresh born Canadian theatre, innocent even of script. Unabashedly exposing its imperfections and incompleteness, yet beautiful in its promise, and because it is truth and it is ours."[76]

Two subcommittees of the VPP, the Victoria Arts and Crafts Council and a choral committee, were formed. The Arts and Crafts committee sponsored a folk art exhibition in mid-February 1974. That summer Theatre Passe Muraille returned to Petrolia. The theatre troupe spent a month studying the community in order to create an original production entitled "Oil," to coincide with the celebration of the town's centennial. The production began with the creation of oil by the "Oil God," and then proceeded at a breathtaking pace through the Oil Springs discovery, the founding of Petrolia, and the emergence of an elite of wealthy oil producers and refiners in Petrolia. Charles Whipp described "Oil" as a "beautiful" play with "some penetrating soliloquies about what Petrolia was really like taken from the minds and mouths of long time Petrolians." Although the characters were not always named, any lifelong resident of the community was able to recognize the characters at once. The play served also as a "memory theatre" – a direct aid for the recall of the past. As Charles Whipp stated in his review: "It was Petrolia, and we understand a little more now why people who live here, like it or not, find it less a place than an experience."[77]

The restoration of Victoria Hall occurred in two phases over a two-year period. It involved a delicate balance between restoring the building to its "original" condition and meeting the contemporary objective of using the building to stage modern theatre productions. Phase I involved adding washrooms, reworking three stairways, and reinforcing the balcony. The objective was to make the building safe and easily accessible. Phase II included decorating the auditorium, installing air conditioning, adding theatre equipment, and converting the fire hall into the main foyer. The total cost of the restorations was nearly $400,000, far exceeding the initial estimates. Lieutenant-Governor Pauline McGibbon officially opened the restored opera house on 1 November 1975.[78]

Further renovations were made to the hall during the 1980s. The entrance was altered in the early 1980s to meet modern-day government regulations that all public facilities be wheelchair accessible. After considerable disagreement and debate between the VPP and the town's Heritage Committee over alternations to the hall's historic appearance, a glass-enclosed foyer was added to avoid losing floor-space within the gallery. The addition, however, significantly altered the outside appearance of the historic building.[79] In 1983 the clock tower was removed and restored with a $250,000 grant from Parks Canada. The original wooden spindles, now rotten, were replaced. Rich Nickels, a machinist at the Oil Well Supply Co., made the new spindles using a flatbed metal lathe normally used to turn oil drilling tools. This was the only lathe available in the area large enough to produce spindles from the 20 cm by 20 cm cedar posts.[80]

Throughout the late 1970s and 1980s a number of well-known Canadian headline acts performed in the Victoria Playhouse, including prima ballerina Karen Kain, guitarist Liona Boyd, the Newfoundland comedy troupe Codco, Second City, the Canadian Opera Company, and Mr Dressup. Early in 1981 the VPP hired consultant John Hill to chart a new course for Victoria Hall. He suggested that the building could better serve the community by opening it up to senior citizens for afternoon card parties and Saturday morning movie matinees for children. Because the building did not depend on any government body other than the town of Petrolia, the VPP had a unique opportunity to use the building as a cornerstone around which the people of the community might gather. "The Shaw and the Stratford Festivals have long gotten away from the people who originally wanted them," Hill stated. "The merchants love them, but the people resent them."[81]

The centennial of Victoria Hall in 1989 was marked by a New Year's levee. Three weeks later, on 25 January 1989, a fire that began in the basement destroyed the building and most of the municipal records housed therein.[82] The recent completion of the restoration project made the loss seem even more tragic. In the interests of saving money, a decision was made not to install a sprinkler system when the building was renovated. The day after the fire, a public meeting was held at the Petrolia Arena in front of a standing-room-only crowd of more than 150 people. Those attending expressed concern for the long-term heritage of the community and supported restoration of Victoria Hall if possible. Mayor Marcel Beaubien emphasized that restoration had to be financially feasible. "We cannot mortgage the future of the Town of Petrolia to restore the building," he stated.[83]

Two architects and two structural engineers toured the remains of Victoria Hall, and upon finding that the exterior brick walls were solid, declared that they were cautiously optimistic that the building could be saved. The town council voted to proceed with the immediate stabilization of the outer shell of the structure, although the council was divided over the issue of restoring the building as opposed to constructing a new municipal building with modern facilities and terminating the connection with the VPP. In March 1989 a group of citizens concerned about the future of the VPP formed the Friends of

Figure 6.7 Fire destroys Victoria Hall, 25 January 1989. Personal collection

VPP.[84] They launched a campaign called "we'll be back," asking for contributions of $1 to help raise $20,000 for the summer theatre season. Meanwhile the VPP refurbished the local Boy Scout Hall and proceeded with its summer production schedule.

Architect Philip Goldsmith of Quadrangle Architects Ltd of Toronto was hired by the town council and a local advisory committee to study the feasibility of restoration, although a decision had not been made to restore the hall. Ultimately, the issue of restoration was determined to a significant extent by the insurance settlement. While the town experts estimated the loss from the fire at between 5 and 5.2 million dollars, the insurers, Zurich Insurance and Scottish and York Insurance, offered only $1,575,000 for the loss.[85] In late November, after several months of negotiation and threats of a lawsuit, the town and the insurance companies settled for the lesser sum of $3.5 million dollars.[86] The citizens of Petrolia, however, were divided over the issue of restoration. Representatives of the Petrolia Heritage Committee and the Friends of VPP appeared before council and presented their case for restoration. Speaking on behalf of the Petrolia Heritage Committee, Mrs Gwen Trone emphasized the historical significance of the building. She stated, "When a town's premier building can evoke such memories and such appreciation, it is not surprising that it also

Figure 6.8 Re-building Victoria Hall. Personal collection

become the focal point of civic pride."[87] Town councillors Roy Ayers and Reeve Ron Snow opposed the restoration scheme, and in February 1990 presented the town council with a petition signed by 410 of the approximately 4,400 residents of Petrolia, calling for the hall to be levelled and replaced with a new single-story building. The petition also called the VPP a burden to taxpayers, and recommended that the theatre group build their own playhouse.[88]

In April 1990 Philip Goldsmith presented four proposals to the town council. Proposal A called for a complete restoration inside and out with the first floor and basement reserved for town offices. The second floor would be reserved for cultural activities, and included the construction of a smaller 420-seat theatre. Proposal B was the same as the first proposal minus the theatre. Proposal C was described as the Cadillac of proposals, costing an estimated $5.6 million. The proposal included a restored hall with a 500-seat theatre and a 316 m^2 addition to the existing structure. The final proposal, proposal D, involved tearing down the hall, salvaging the clock tower, and building new offices around it. By a margin of 6–3, town council voted to restore the building according to Proposal C, but with a smaller 425-seat theatre. Mayor Beaubien referred to the importance of the town's Victorian oil heritage, and remarked that "Victoria Hall is the key or anchor of this."[89] A nine-member fundraising committee composed of Mayor Beaubien, Deputy Reeve Isber, Councillor Ross O'Hara, and six other prominent citizens was organized with a goal of raising $750,000 towards the restoration.

Almost two years after the fire, the reconstruction of Victoria Hall began on 19 November 1990. Members of the local heritage committee sifted through the rubble in an effort to save artifacts from the original building, particularly the ornate iron panels from the cast iron balcony and the two opera boxes in the original building. The restoration was completed in September 1992, at a cost of $6.5 million. The insurance settlement, funds raised locally, and federal and provincial grants covered the cost of restoration. The building was restored in a manner that architect Philip Goldsmith described as sympathetic to the original. A concerted effort was made to mediate between traditionalist and modernist goals. A modern building with computer controlled heating and ventilation was constructed inside the original shell. The exterior was restored to its original style before the glass-enclosed foyer was added in the 1970s. The front and back steps were returned to their original design. The upper floor of the building houses the new theatre, offices, and space for storage. A stereo system and a system for the hearing impaired were installed in the theatre. The middle floor contains the council chamber with seating for 60 people, meeting rooms, and a large theatre lobby with a kitchen and a bar. The ground floor was expanded to hold the municipal offices, plus the washrooms for the staff and the theatre. A terrace was added to the exterior of the building off the lobby of the theatre and council chambers. The original bell, which was made of brass and cast in New York in 1876, was cracked in the fire and could not be used in the

Figure 6.9 The re-opening of Victoria Hall, September 1992. Personal collection

restored building. Another bell was found in a Pennsylvania warehouse, and purchased for $10,000. Albert Scarsbrook, a coffee and tea merchant, donated the Union Jack that flew at Victoria Hall around the period of World War I. His daughter, Helen Corey, donated a new Canadian flag for the reopening of the restored building.[90]

A week of gala festivities was held from 20–27 September 1992 to celebrate the reopening of Victoria Hall. The *Advertiser-Topic* described the hall as a "modern version of the Greek's phoenix." When the flames had died, "a new bird, one as glorious and as beautiful as the first would be born from those remains." "The legacy has been rebuilt," Mayor Beaubien remarked during the ceremonies marking the reopening of the hall. Victoria Hall has been, and remains, a symbol of place around which public memory was constructed in Petrolia. The key reason to restore the building, the *Advertiser-Topic* suggested, was that the building has meaning, and "countless memories have been made in Victoria Hall." Some of the memories of long-time residents of Petrolia were published in a special commemorative issue of the *Advertiser-Topic* published on 16 September 1992, entitled "Rebuilding the Legacy." Helen Corey remembered singing a solo in the hall when she was only five or six years old. She recalled standing on the wings, "scared to death" until someone literally pushed her onto

Figure 6.10 Ben Skinner's mural: "Canada's Victorian Oil Town." Personal collection

the stage. Another long-time resident, Randolph Thomson, recalled that during the 1930s people flocked to Victoria Hall to see the boxing and wrestling matches organized by promoter Jack Fraser. For long-time residents of Petrolia, Victoria Hall evokes memories.[91]

Consciousness-raising about heritage was successful in Petrolia. The debate over whether to restore Victoria Hall after the 1989 fire served to mobilize concerns about community identity. The process of restoring Victoria Hall also reveals that saving a public past in any town is a political as well as a historical and cultural process. Various organizations, ranging from the municipal council to community voluntary groups such as the Friends of VPP, worked alongside one another to bring about the restoration of Victoria Hall. This is not to say that disagreements did not occur between groups, first over whether it was desirable and feasible to even attempt restoration, and later over how to restore the building.

The restoration of Victoria Hall and the development of the historical oil field were two components of the project for revitalizing Petrolia. The third component is the ongoing conservation of the Victorian downtown façade. Petrolia's downtown deteriorated in the 1960s when the town's primary retail function was lost as residents turned to the large shopping malls that emerged in Sarnia and London. As in other Ontario towns, much of Petrolia's architectural heri-

Figure 6.11 Street signs symbolize Petrolia's oil heritage. Personal collection

tage is found on the main street. Since the 1970s preservation of that heritage has depended on the ability of owners to maintain or restore their buildings. The Petrolia Heritage Committee and the Petrolia Local Architectural Conservation Advisory Committee, formed in 1974 under the Ontario Heritage Act, have promoted the preservation of the town's architectural heritage. The main disincentive for the owners of these historic properties has been that renovation has often resulted in substantial increases in property taxes.

In 1996 the town budgeted $35,000 to start a façade restoration program for owners of commercial buildings along the main street. The program is funded by the town and administered by the Petrolia Heritage Committee. The town provides 50 per cent of the costs to property owners who undertake approved restoration work to the façade of a building in the designated Heritage District of Petrolia. The heritage committee has assisted with painting, window replacement, structural work, and figurative metal work with sensitivity to the original Victorian design and materials.[92] The town council and the heritage committee have used the heritage to create a "visitable" place which they market as "Canada's Victorian Oil Town." Tourists are invited to explore the Victorian architecture with the promise that "the opulence of the oil barons will delight and amaze."

Public spaces have been made over so that they actively invite the attention of visitors. Emblems and insignia have been used to create the townscape. Cultural display is everywhere in Petrolia. Lampposts are constructed as replicas of oil derricks. Street signs are also emblazoned with oil derricks as heritage has blended with the market forces of tourism. Signposts portray the opulence of the oil barons and Victoria Hall is the centre of the attraction. Banners attached to lampposts promote Petrolia as "Canada's Victorian Oil Town." Artists and designers have been active in the work of making a unique place that attracts the tourist gaze. In 1997 local artist Ben Skinner painted two murals in the downtown area: the first an oil field scene with the words "Canada's Victorian Oil Town," and the other a collage of Petrolia buildings with Victorian architecture.[93] Cultural display in Petrolia is aimed at getting visitors to spend money in retail establishments on the main street, attend the highly successful theatre program staged at Victoria Hall, and experience the living museum at the Petrolia Discovery.

Conclusion

On 3 August 2002 fire swept through the Fairbank mansion heavily damaging the attic and slate roof. Built by J.H. Fairbank, and completed in 1891, the house befitted his status as an "oil baron" and the "Father of Petrolia." Located at the head of the rows of Petrolia's business blocks at the east end overlooking the town, the mansion, named Sunnyside by the family, was impressive in its heyday, a marvel of late Victorian architecture and craftsmanship. Massive, yet elegant in its appearance, the exterior of the mansion was built of red brick of Ohio clay. Each brick was wrapped in wax paper before shipment. Most of the timbers and hardwoods used in the trim were cut from the Fairbank farms and dried for a year prior to construction. The fireplaces were constructed of ceramic marble. Sunnyside had twenty-two rooms, including a ballroom, a reception room, a library, a billiard room, and eight bedrooms, each with its own dressing room and bathroom. A massive wooden staircase, capable of holding five abreast, led up to the third storey ballroom where famous orchestras entertained the elite of Petrolia society: the Coreys, the Lanceys, the Kerrs, the Moncrieffs, and the Nobles among others. Earl Grey, Lord Lascelles, and Pauline Johnston stayed at the Fairbank mansion during their visits to Petrolia.

Dr Charles O. Fairbank's widow was the last member of the Fairbank family to live in the mansion. After she died in 1956, the mansion was used as a temporary home for the local Anglican congregation following the 1957 fire that destroyed Christ Church. Subsequently, two

The Fairbank mansion following the 3 August 2002 fire.
Personal collection

women from Sarnia, Jean Scott and Mrs Frank Ryan leased the mansion and turned it into a rest home. In 1963 it was transformed once again into a residential adult education centre named Fairbank House. The program stressed the development of skills in the arts and crafts, management and supervisory training, trade union affairs, rural and farm concerns, youth leadership, and municipal government training. At the end of 1965 Fairbank House was operating with a deficit of $1,500 and a low enrolment. "Opulent but obsolete," the Fairbank mansion faced an uncertain future, and Charles Fairbank, the grandson of J.H. Fairbank, hinted that demolition was a possibility.[1] In 1967 Charles Fairbank sold the mansion to R.J. Burnie of Sarnia and most of the contents were sold by auction.

At the time of the August 2002 fire the Burnie family was still the owner of the mansion, although it was no longer a single family dwelling, having long since been divided into five apartment units. Ronald and Jessie Burnie rejected an offer made by the town council in July 1983 of a six-month option to purchase the mansion for $90,000.[2] The building has never been designated a heritage building under the 1975 Ontario Heritage Act. Once again the Fairbank mansion faces an uncertain future. The building is privately owned and there are no guarantees that the owner will rebuild, although a new roof has since been constructed. Estimates of costs to repair the mansion range from $2 million to $3.5 million. There are also concerns about the architectural integrity of any restoration, as the mansion was an example of the finest that nineteenth-century craftsmanship had to offer. Some residents have suggested that the town council should assist in the restoration of the mansion, referring to the commitment the council showed in resorting Victoria Hall after that building was devastated by fire in 1989. In endorsing the restoration of the mansion, Charles Fairbank, the great-grandson of J.H. Fairbank, states that "the building is a historically significant building that's 'symbolic' of Petrolia's oil boom past."[3] A community volunteer group calling itself the "Sunnyside Committee" recently applied to the Ontario government for a heritage designation for the mansion. The old question that had been asked for years cropped up once again: Why wasn't something being done to protect this building?

Not long after fire ravaged the mansion, freelance journalist Kelly-Jo Moxam wrote, "The Fairbank mansion is simply a part of the collective psyche of the town."[4] At risk if the Fairbank mansion is demolished is a community identity constructed around the representation of Petrolia as "Canada's Victorian Oil Town." Another devastating blow to the preservation of Petrolia's legacy as an oil-boom town occurred on 11 August 2004 when a fire destroyed part of the downtown façade.[5]

Throughout this book, I have argued that the transformation of Petrolia from a resource town into a respectable Victorian community occurred in a dialectic between the local, colonial context and Empire, in which the local oil industry effected a formative role. This study has been informed by theories of place as an amalgam of social and cultural processes rather than merely a stage upon which historical actors performed.

Major transformations in the organization of colonial governance were initiated beginning in the early decades of the nineteenth century in the settler colony of Upper Canada and a shift towards colony-centred knowledge production occurred. European practices of geographical reconnaissance made the colonies amenable to colonization. Township surveying and land registry were connected in Upper Canada as colonial administrators transformed wilderness into bounded property, thus legalizing imperial claims to territory but dispossessing native peoples. Influenced by the writings of post-colonial scholars, I argued that colonialism was never a uniform or coherent imposition, but a practically mediated set of social relationships. Imperial discourses were bent from colonial vantage points at the local level where most settlers, squatters, and land speculators had dealings with the ruling apparatus of the state. Squatters who stripped the land for timber and engaged in the process of farm building did not always adhere to imperial models of affiliation, loyalty, and dependence. With the oil boom of the early 1860s land speculation in Enniskillen Township was rampant. The attempts by speculators to profit from their investment in land was part-and-parcel of the normal workings of capitalism, but ran counter to the idealized world of the settler-citizen imagined by colonial administrators.

For historians of colonial land policies, the "scale of operation" approach is not an adequate indicator of the extent of land speculation in Upper Canada, as it fails to disclose the motivations of land dealers and consider local variations in social, economic, and geographical conditions. As the case study of Enniskillen Township revealed, most of the speculative activity occurred in a limited geographical area on small parcels of land where substantial quantities of oil were known to exist. Profit was always the primary motivation behind speculation in oil lands, and there was a tendency for quick turnover rather than long-term investment. Thus, the "scale of operation" approach would not have revealed the true extent of land speculation in Enniskillen Township during the oil boom of the early 1860s. The analysis of land holding in Enniskillen Township also points to the continued relevance of micro-histories so that the effects of colonial land granting policies might be more fully understood at the level where they were played out.

The power of science mesmerized the Victorians. Science and culture were linked in the eyes of the Victorians as they defined knowledge and ordered nature. A culture based on natural science and progress was formed in the Enniskillen oil fields around the middle of the nineteenth century. The construction of scientific knowledge was worked out locally. In carrying out the geological survey of Enniskillen Township professional scientists relied on the observations of local settlers and practical men in the field. Science was enmeshed in the warp and woof of commercialism, colonization, and culture as professional geologists, practical men, and oil smellers sought to determine the location and the extent of local oil resources. The colonies, therefore, were not merely an object of study and a source of data for scientists in Britain. Rather, the production of scientific knowledge occurred in a dialectical relationship between the colony and Empire. In studying the construction of scientific knowledge in the Enniskillen oil fields, I have argued that the work undertaken in the colony must be understood in its own terms, and in relation to scientific traditions developed in Canada.

Following the collapse of the first oil boom at Oil Springs in 1863, the focus of oil production shifted to Bear Creek. The oil industry provided the expression and precise definition of communal boundaries and shaped class identities in the fledgling nineteenth-century resource town of Petrolia. This, however, was not solely, or exclusively, an economic process. In Petrolia communal boundaries were constructed in socio-cultural space and were maintained through the discourses of everyday life. "Insiders" shared access to a common discourse built around the vagaries of the oil industry that was unfamiliar to those outside the community, to whom access was denied. The historical study of Petrolia suggests that building a community identity in nineteenth-century Ontario towns was a process that varied from town to town, and it was interwoven also with national and international trading patterns. As the local oil economy of Petrolia was integrated more extensively into national and international markets beginning in the early 1870s, the material conditions were present for the emergence of a middle class. The predominance of small oil producers and the capriciousness of the markets made for a highly competitive and volatile industry and class lines were fluid.

A middle class of prosperous "oil barons," merchants, manufacturers, and professionals transformed Petrolia from a crude shantytown into a respectable Victorian community; one that was orderly, regulated, learned, prosperous, and "civilized." Public space, as the analysis of associational culture in nineteenth-century Petrolia revealed, was constructed as the terrain of Anglo-Protestant middle-class men. Male associational life in nineteenth-century Petrolia operated in a two-sided cultural context where working-class men could be pulled in contradictory directions. Fraternal orders encouraged cross-class mutualism and provided the ideals of morality that allowed the Knights of Labor to flourish in Petrolia for a brief period in the 1880s. Fraternal hierarchies, however, also resulted in the formation of elite groups, who gained valuable experience in community leadership, and who in turn shaped civic life. Civic celebrations and amateur sports were organized by the male middle class, and served to legitimize their ideals of manly respectability.

Gender shaped the creation of public space in Petrolia during the "boom town" era. The concept of separate spheres, as women's historians have long argued, was a powerful prescriptive ideology during the nineteenth century and integral to the shaping of middle-class identity. The home was associated with a particular middle-class ideal of feminine domesticity based on the family, while the public sphere was constructed as the province of middle-class men. Although middle-class men dominated public space in nineteenth-century Petrolia, the role of middle-class women was vital to the shaping of a culture of refined sociability. Women's involvement in mixed-sexed literary societies, for example, allowed them to hone their writing and public speaking skills, and to gain a foothold into the civic affairs of the town. Both middle-class women and working-class women selectively used separate spheres ideology to make claims in public based on an ideal of respectability rooted in domesticity. Although the concept of separate spheres remains a useful explanatory framework for historians, I would concur with those historians who have suggested that the ideology of separate spheres cannot provide the basis for a universal sisterhood or generalized gender consciousness.[6]

During periods when the markets for oil declined local drillers were left unemployed. In December 1873 the first crew of foreign drillers left Petrolia for the Dutch colony of Tjibodas in West Java. Following the takeover of Imperial Oil by the U.S.-owned Standard Oil in 1898, and the demise of local oil supplies a decade later, the exodus of oil drillers from Enniskillen Township for foreign fields escalated.

The foreign drillers from Enniskillen Township, many from the town of Petrolia, provided the skilled labour and technical expertise necessary for the development of the global oil industry. The travel writings produced by the foreign drillers were part of the process by which European imperialism was produced for home consumption. The foreign drillers constructed a cultural space in which they disavowed the creative agency of non-white native Others using prior constructions of racial and gender hierarchies that furthered the project of European imperialism. The mapping of colonial masculinities by foreign drillers for their readers back home in Enniskillen Township occurred in a dialectic of "home" and "away." Labour and domesticity were means by which the mutually constitutive processes of class formation and racialization occurred in the colonial oil fields. An argument was made for an appreciation of the peculiarities of the Canadian oil drillers both as British subjects of a white settler colony and as colonizers in the colonies of oil exploitation. In doing so, I have attempted to respond to recent calls by some historians to put the imperial experience back to the writing of nineteenth-century Canadian history.

The local economy of Enniskillen Township benefitted from foreign drilling. While working abroad, most of the foreign drillers sent money home regularly to support their wives and families. Local manufacturers and retailers profited from the global business in colonial oil development. Employment abroad resulted in elevation into the ranks of the middle class for a substantial number of foreign drillers. While a handful became wealthy men and never returned to Enniskillen Township, others attained supervisory or managerial positions with large international oil companies. Some foreign drillers were able to purchase farms, start small businesses, or purchase local oil properties upon returning to Enniskillen Township.

Attempts at industrial reorientation were made in Petrolia in the aftermath of the relocation of the Imperial Oil Company's refining operations to Sarnia in 1898. The exodus of skilled oil drillers to foreign fields also continued into the twentieth century. The building of a public memory around Petrolia's history as an "oil boom" town has been integral to the re-assertion of community identity. The Old Boys' celebrations in 1908, 1925, and 1946 were used to highlight the role of public memory and place in shaping community identity. Public memory was constituted from shared childhood memories of local pleasure spots.

Beginning in the 1970s a new economic and commercial base for the town was built around the commercialization of heritage to promote tourism. Petrolia is now marketed as "Canada's Victorian Oil Town." The economic revitalization of Petrolia around tourism emphasizes the town's past as a nineteenth-century oil-boom town. The "oil barons" symbolize the wealth and status of Petrolia as a respectable middle-class family town. The foreign drillers have been used to highlight the importance of Petrolia to the development of the global oil industry. The preservation of the historic oil field as a living museum, the conservation of the downtown Victorian façade, and the restoration of Victoria Hall as a venue for summer theatre have been targeted as key projects in this redevelopment of Petrolia around tourism.

Throughout this book an argument has been made for studying "colony" and "Empire" and "home" and "away" as a dialectical rather than dichotomous relationship. Far too often the work of the local historian has been dismissed as simply parochial. I have suggested that local micro-studies remain relevant for considering how power-knowledge relationships were mediated from colonial vantage points. Emphasis was placed on the social and cultural processes through which one small Ontario town came to be remembered, and more recently marketed for tourism as "Canada's Victorian Oil Town." As one life- long resident of Petrolia stated in referring to the importance of this particular small town to the development of global oil industry: "we were famous around the world; we were really something."[7]

Notes

INTRODUCTION

1 Radforth, *Bush Workers and Bosses*; McIntosh, *Boys in the Pits*; Heron, *Working in Steel*; Bray and Thomson, *At the End of the Shift.*

2 Buchner, "Whatever happened to the British Empire," 3–32.

3 Canada, *Indian Treaties and Surrenders*, vol. I, 58, 71, 244, 246; Surtees, "Indian Land Cessions in Upper Canada, 1815–1830," 65–83; Elford, *Canada West's Last Frontier*, 45; Telford, "Under the Earth," 65–79; *Sarnia Observer*, 12 May 1977.

4 Clayton, *Islands of Truth*; Stasiulis and Yuval-Davis, *Unsettling Settler Societies*; Edney, *Mapping and Empire*; Clayton, "Absence, Memory, and Geography," 65–79; Gregory, "Imaginative geographies," 447–85.

5 Curtis, "Official Documentary Systems and Colonial Government," 389–417.

6 Zeller, *Inventing Canada*, 5, 184–5; Clarke and Finnegan, "Colonial survey records and the vegetation of Essex County, Ontario," 199–238.

7 Clarke, *Land, Power, and Economics on the Frontier of Upper Canada*; Shannon, "Brokers, Land Bankers, and 'Birds of Evil Omen';" Widdis, "A Perspective on Land Tenure in Upper Canada."

8 Clayton, "Absence, Memory, and Geography," 65–79.

9 Secord, "Science in the Pub," 269–315; Secord, "Corresponding interests," 283–408; Knell, *The Culture of English Geology*, 3–27; Torrens, William Edmond Logan's Geological Apprenticeship in Britain 1831–1842," 97–110; Lucier, "A Plea for Applied Geology," 283–318.

10 White, *Geology, Oil Fields and Minerals of Canada West.*

11 Shields, *Places on the Margin*; Soja, *Postmodern Geographies*; Cohen, *Symbolising boundaries*; Thrift and Williams, *Class and space*; Lefebvre, *The Production of Space*; Certeau, *The Practice of Everyday Life*; Massey, *Space, Place, and Gender.*

12 Ryan, *Cradle of the Middle Class*; Davidoff and Hall, *Family Fortunes.*

13 See, for example, Kealey, "Orangemen and the Corporation," 41–86.

14 Ryan, *Women in Public,* 19–35; Morgan, *Public Men and Virtuous Women,* 183–214.

15 The term "foreign drillers" was widely used by the residents of Enniskillen Township to refer to local drillers who travelled abroad.

16 Spurr, *The Rhetoric of Empire*; Hall, "Going-a-Trolloping," 180–99; Dubinsky, *The Second Greatest Disappointment*; Dubinsky, "Local Colour," 67–79; Stoler, "Rethinking Colonial Categories," 134–61.

17 Pratt, *Imperial Eyes,* 4–7.

18 Burton, "Rules of Thumb," 483–500; McClintock, *Imperial Leather*; Marks, "History, the Nation and Empire," 111–19; Mackenzie, *Imperialism and Popular Culture*; Tabili, '*We Ask for British Justice*'; Burton, "Who Needs the Nation," 228–9; Hall, *Cultures of Empire.*

19 Ryan, *Women in Public*; Davis, *Parades and Power*; Radforth, *Royal Spectacle;* Walden, *Becoming modern in Toronto*; Gordon, *Making Public Pasts*; Nelles, *The Art of Nation-Building.*

20 Halbwachs, *The Collective Memory,* 140.

21 Hayden, *The Power of Place,* 2–45.

22 Samuel, *Theatres of Memory, Vol. 1,* ix-17; Thelen, "Memory and American History," 1122–3; Tallentire, "Strategies of Memory," 197–212; Walsh and High, "Rethinking the Concept of Community," 255–73.

23 Samuel, *Theatres of Memory*; Lowenthal, *The Past is a Foreign Country*; Lowenthal, *Possessed by the Past*; Walsh, *The Representation of the Past;* Harvey, *The Condition of Postmodernity*; Dicks, *Culture on Display*; Dicks, *Heritage, Place and Community*; Osborne, "Landscapes, Memory, Monuments, and Commemoration," 39–77; Wallace, *Mickey Mouse History and Other Essays on American Memory*; McKay, *The Quest of the Folk.*

24 Harvey, *The Condition of Postmodernity,* 62–3, 85–7; Walsh, *The Representation of the Past,* 1–37.

25 Dicks, *Culture on Display,* 13, 29–30; Urry, *The Tourist Gaze,* 1–3.

CHAPTER ONE

1 Widdis, "Motivation and Scale," 338–51; Shannon, "Brokers, Land Bankers, and 'Birds of Evil Omen';" Clarke, *Land, Power, and Economics on the Frontier of Upper Canada,* 304–12.

2 Patterson, *Land Settlement in Upper Canada 1783–1840*, 32.
3 Clarke, "The Role of Political Position and Family and Economic Linkage in Land Speculation in the Western District of Upper Canada, 1788–1815," 18–34; Brunger, "A spatial analysis of individual settlement in Southern London District, Upper Canada, 1800–1836"; Gagan, "Property and Interest," 63–70.
4 Widdis, "Motivation and Scale," 338–51; Widdis, "A Perspective on Land Tenure in Upper Canada"; Shannon, "Brokers, Land Bankers, and 'Birds of Evil Omen'."
5 S.P. Hurd to Lewis Burwell, 13 July 1832, RG1, Archives of Ontario.
6 Walsh, "Property, Surveyors, and State Formation"; Richards, "Archive and Utopia," 104–35.
7 Lewis Burwell to S.P. Hurd, Esq., Surveyor General, 17 July 1832, RG1, A-I-1, Vol. 16, MS 30, Reel 2, Archives of Ontario; Survey Diary of Lewis Burwell, 14 August 1832, RG1, Series CB-1, Box 9, MS 924, Reel 6, Archives of Ontario. Mahlon Burwell was compensated for much of his work in land. In 1830–31 Mahlon and his brother John laid out a plot for the village of Port Burwell on land he was granted at the mouth of Otter Creek for his work in surveying Bayham Township. See Clarke, *Land, Power, and Economics on the Frontier of Upper Canada*, 13; *Dictionary of Canadian Biography*, VII, 125–8; Ladell, *They Left Their Mark*, 112; 121–2.
8 Ladell, *They Left Their Mark*, 118.
9 Bureau of Archives, 1906, 384, cited in Gentilcore and Donkin, *Land Surveys of Southern Ontario*, 13.
10 Ibid., 15–17.
11 Edney, *Mapping an Empire*, 77–8; Pratt, *Imperial Eyes.*
12 Lewis Burwell to S.P. Hurd, 9 January 1833, RG 1, A-I-1, Vol. 16, MS 30, Reel 2. Survey Diary of Lewis Burwell, 1 September 1832.
13 Zeller, *Inventing Canada*, 5.
14 Clarke and Finnegan, "Colonial survey records and the vegetation of Essex County, Ontario," 119–38; Zeller, *Inventing Canada*, 184–5.
15 Edney, *Mapping an Empire*, 77–8.
16 Survey Diary of Lewis Burwell, RG 1, Series CB-1, Box 9, MS 924, Reel 6, Archives of Ontario.
17 Lewis Burwell, Enniskillen Field Book.
18 Enniskillen Field Book; Lewis Burwell to S.P. Hurd, 9 January 1833, RG1, A-I-1, Vol. 16, MS 30, Reel 2, Archives of Ontario.
19 Kelly, "The Artificial Drainage of Land in Nineteenth-Century Southern Ontario," 279–98.
20 Johnson, *Becoming Prominent*, 54.

21 *Dictionary of Canadian Biography*, VII, 335–40; *Dictionary of Canadian Biography*, XI, 125–31; *Dictionary of Canadian Biography*, V, 411–12; Palmer, *A Culture in Conflict*, 100–7.
22 Clarke, *Land, Power, and Economics*, 184–6.
23 Under the proclamation of 16 July 1792, Upper Canada was divided into nineteen districts. Essex and Kent counties were part of the Western District. Initially, the townships of Moore, Sarnia, Plympton, Enniskillen, Warwick, Brooke, and Bosanquet were attached to the County of Kent. It was not until much later, in 1835, that the surveys of the ten townships comprising Lambton County were completed. In 1852 Lambton became a provisional county. The union with Essex was formally dismantled on 30 September 1853, and Lambton became a fully independent county.
24 John Ward to the Honourable J.H. Price, 6 September 1851, Enniskillen Township Papers.
25 Sir John Colborne was concerned about the defence of the western frontier. He visited St Clair in 1835 and gave orders for a survey of a direct route from London, where he planned to establish a military depot, to a point on Lake Huron. The Egremont Road was designed to link London with the county town of Errol. See Lauriston, *Lambton's Hundred Years*, 72–4; *Belden's Illustrated Historical Atlas of the County of Lambton*, 15.
26 Elford, *Canada West's Last Frontier*, 45; Canada, Census of 1861, Enniskillen Township.
27 Johnson, "Land Policy, Population Growth, and Social Structure," 41–61; Russell, *Attitudes to Social Structure*; Joy Parr, "Hired Men," 91–103; Bitterman, "The Hierarchy of the Soil," 33–55; Samson, "Industry and Improvement."
28 RG 1, C-IV Township Papers, Enniskillen Township, MS 658, Reel 132, Archives of Ontario.
29 Henry Clarke to the Commissioner of Crown Lands, n.d., Enniskillen Township Papers.
30 General Abstract of the Collection & Assessment Rolls for the Western District for the year 1843, Lambton County Library, Wyoming, Ontario, MS 577, Reel 1; Sutherland Bros, *County of Lambton Gazetteer, and General Business Directory for 1864–5*, 61.
31 Canada, Census of 1861, Enniskillen mss.
32 Clarke, *Land, Power, and Economics on the Frontier of Upper Canada*, 271–86; Gagan, "The Security of Land," 135–53.
33 George Wright and John Gaus, 15 May 1855, Township Papers.
34 Township Papers.
35 Canada, Census of 1861, Enniskillen mss.
36 *Sarnia Observer*, 1 February 1861.

37 *Sarnia Observer*, 15 March 1861; 16 March 1861; 31 May 1861.

38 *Sarnia Observer*, 22 February 1861.

39 *Sarnia Observer*, 22 March 1861.

40 Johnson, "Land Policy, Population Growth, and Social Structure in the Home District, 1793–1851," 41–61; Clarke, *Land, Power, and Economics*, 296.

41 Clarke, "The Role of Political Position and Family and Economic Linkage in Land Speculation," 18–34; Brunger, "A spatial analysis"; Gagan, "Property and Interest," 63–70.

42 Widdis, "Motivation and Scale," 338–51; Shannon, "Brokers, Land Bankers, and 'Birds of Evil Omen'."

43 Clarke, *Land, Power, and Economics*, 305–7.

44 At the height of the oil boom in the early 1860s property changed hands quickly with a handshake. Many of these transactions were never recorded and a number of disputed claims made their way to Chancery Court.

45 McCalla, *Planting the Province*, 141–2. See also Clarke, "The Role of Political Position and Family," 18–34; Johnson, *Becoming Prominent*, 37–60.

46 Johnson, *Becoming Prominent*, 179–80; *Dictionary of Canadian Biography*, X, 124–9.

47 Bailey, *Dictionary of Hamilton Biography*, I; *Dictionary of Canadian Biography*, XI, 929–30.

48 *Sarnia Observer*, 30 December 1858.

49 *Hamilton City Directory, 1862–63*, 16.

50 See Enniskillen Abstract Index A, 1861, Sub-division, for the plan of Williams, Anthony, and Bush, mfm copy, Archives of Ontario. The sub-division covered four hundred acres on lots 16 and 18, concession one, and lots 16 and 17, concession 2.

51 *Globe*, 2 September 1871.

52 *Globe*, 25 April 1865.

53 Andrew Elliott to George Brown, 7 April 1865; George Brown to Alexander Campbell, 23 January 1865; Memorandum from A. Campbell, 26 April 1865; George S. McPherson to Andrew Elliott, 19 April 1865; 22 April 1865, Enniskillen Township Papers. On 30 June 1865, the *Sarnia Observer* suggested that collusion had occurred between the subordinates of the Crown Lands Department and the purchasers of the lands. By an arrangement between twenty or thirty land speculators, each was allowed to bid without competition. The *Observer* charged that prominent friends of George Brown were involved in the transaction. See *Sarnia Observer*, 30 June 1865.

54 Clarke, *Land, Power, and Economics*, 301–2.

55 McEvoy & Co., *Gazetteer and Directory of the Counties of Kent, Lambton, and Essex, 1866–7*, 269.

56 *Belden's Illustrated Historical Atlas of the County of Lambton,* 12; Harkness, "Makers of Oil History, 1850–1880," unpublished ms, Leddy Library, University of Windsor, 10–11; May, *Hard Oiler,* 29–30; Morritt, *Rivers of Oil,* 17–23.

57 Chambers, *Married Women and Property Law in Victorian Ontario,* 66–7. The Court did not authorize Mrs Tripp to take control of the estate herself. Because divorce was unavailable to Mrs Tripp, there was always the possibility that her husband would return and enforce his marital rights. Vice-Chancellor Spragge ordered that new trustees be appointed to serve the interests of Mrs Tripp; that she be paid one-half the rents and profits from the estate; and that the other half be paid into the Court in trust for her husband.

58 *Sarnia Observer,* 2 November 1866.

59 *Globe,* 2 September 1861, 6 September 1861; 12 September 1861.

60 Instrument No. 1289, Land Register Volume C, Enniskillen Township.

61 *Sarnia Observer,* 5 April 1861.

62 No. 1086, Copy Book Register B, Enniskillen Township; Enniskillen Abstract, Index Volume A.

63 Reprinted in the *Sarnia Observer,* 14 February 1862. Shaw also patented a still for refining oil, and operated a refinery until it was destroyed by fire in May 1862. On 11 February 1863, Shaw fell into his well and drowned. See *Globe,* 14 February 1863; 20 February 1863.

64 Phelps, "Foundations of the Canadian Oil Industry, 1850–1866," 160.

65 Newell, *Technology on the Frontier,* 121–7.

66 *Globe,* 14 April 1865.

67 Lambton Oil Sales, Memorandum of Instructions, Crown Land Papers, RG 1, G-1, Volume 8, Archives of Ontario.

68 Edgar, *A Manual for Oil Men and Dealers in Land,* viii, 41.

69 McBride, *Poems Satirical & Sentimental,* 111.

CHAPTER TWO

1 Edney, *Mapping an Empire.*

2 Secord, *Victorian Sensation,* 42–56.

3 Lightman, *Victorian Science in Context*; Teich and Young, *Changing Perspectives in the History of Science.*

4 Secord, "Science in the Pub," 269–315; Secord, "Corresponding interests," 383–408; Knell, *The Culture of English Geology, 1815–1851,* 3–27.

5 Torrens, "William Edmond Logan's Geological Apprenticeship in Britain 1831–1842," 97–110; Lucier, "A Plea for Applied Geology," 283–318; Turner, "Practicing Science: An Introduction," in Lightman, ed., *Victorian Science in Context,* 286.

6 Zaslow, *Reading the Rocks*. For an overview of early trends in the social history of geology see Jordanova and Porter, *Images of the Earth*, v-xvi.
7 Zeller, *Inventing Canada*, 48–9.
8 Eagan, "The Canadian Geological Survey," 99–106.
9 Stafford, *Scientist of empire*; Secord, "The Geological Survey of Great Britain as a Research School, 1839–1855," 223–75.
10 Hunt, "Doing Science in a Global Empire" in Lightman, *Victorian Science in Context*, 312–13.
11 Cooper and Stoler, *Tensions of Empire*, 1.
12 Gale, *The Wonder of the Nineteenth Century!*; Henry, *The Early and Later History of Petroleum*, 10–11.
13 Cruikshank, *The Correspondence of Lieut. Governor John Graves Simcoe*, 290.
14 Simcoe, *The Diary of Mrs. John Graves Simcoe*, 155.
15 Howison, *Sketches of Upper Canada*, 194.
16 Pickering, *Enquiries of an Emigrant*, 122.
17 Knell, *The Culture of English Geology*, 309.
18 Owen, *Trek of the Oil Finders*, 15–17.
19 The biographical information on William Logan is drawn from the following sources: Harrington, "Sir William Logan," 1–13; Harrington, *Life of Sir William E. Logan Kt*; Alcock, *A Century in the History of the Geological Survey of Canada*; Zaslow, *Reading the Rocks*, 3–61; Zeller, *Inventing Canada*, 13–50; Winder, "Sir William Edmond Logan," *Dictionary of Canadian Biography*, X, 444–9.
20 Torrens, "William Edmond Logan's Geological Apprenticeship," 98.
21 Harrington, *Life of Sir William E. Logan*, 49.
22 Porter, "The Industrial Revolution and the Rise of the Science of Geology," in Teich and Young, *Changing Perspectives in the History of Science*, 320–43; Knell, *The Culture of English Geology*, xvii, 5–7, Secord, "Science in the Pub," 270–1.
23 Porter, "The Industrial Revolution and the Rise of the Science of Geology," 326–7.
24 Harrington, *Life of Sir William E. Logan*, 159, 222; Ladell, *They Left Their Mark*, 132.
25 Secord, *Controversy in Victorian Geology*.
26 Eagan, "The Canadian Geological Survey," 99–106; Zaslow, *Reading the Rocks*, 28–30; 41–3.
27 Duchesne, "Thomas Sterry Hunt," *Dictionary of Canadian Biography*, Vol. XII, 457–58; Brock, "Chemical geology or geological chemistry," in Jordanova and Porter, *Images of the Earth*, 150–1.
28 Legislative Assembly, *Report of the Progress of the Geological Survey of Canada for the Year 1849–50*, 99.

29 Knell, *The Culture of English Geology*, 6.
30 *Report of the Progress of the Geological Survey of Canada for the Year 1849–50*, 99.
31 Legislative Assembly, *Report of the Progress of the Geological Survey of Canada for the Year 1850–51*, 29–30, 33.
32 Alexander Murray, Field Book, Enniskillen Township, RG 45, Vol. 168, Notebook 2698, National Archives of Canada.
33 Legislative Assembly, *Report of the Progress of the Geological Survey of Canada, 1851–52*, 90–1. A chain is a unit of measurement: one chain equals four rods or 66 feet.
34 Morritt, *Rivers of Oil*, 17–21.
35 A copy of Antisell's report, dated 19 February 1853, is contained in the Tripp Family File, Lambton Room, Lambton County Library, Wyoming, Ontario.
36 Petition of Charles N. Tripp and others for the Formation of the International Exploring, Mining and Smelting Company, 26 October 1854. Copy Oil Museum of Canada, Oil Springs, Ontario.
37 Ewing, "The History of Imperial Oil Limited," unpublished ms, Harvard Business School 1951, 23.
38 Newell, *Technology on the Frontier*, 121.
39 *Sarnia Observer*, 5 August 1858. The account was originally published in the *Woodstock Sentinel*. Considerable nationalistic debate has taken place over whether James Miller Williams was the first man to *drill* successfully for oil in North America. Popular U.S. sentiment attributes this distinction to Edwin L. Drake, who discovered oil near Titusville, Pennsylvania, on 27 August 1859 after drilling 69 feet. The debate resurfaced in 1960 in a series of letters between Colonel R.B. Harkness, Ontario oil commissioner and amateur oil historian, and Ernest C. Miller, vice-president of West Penn Oil Company of Warren, Pennsylvania. The account from the *Sarnia Observer* suggests that Williams' discovery pre-dates Drake's find by slightly more than a year. In an 1860 U.S. publication, entitled *The Wonder of the Nineteenth Century! Rock Oil, In Pennsylvania and Elsewhere*, intended to satisfy public demand for more information about petroleum, Thomas Gale acknowledged Williams' prior discovery: "Williams & Co's well is 49 feet deep, 7 by 9 feet square, cribbed with small logs, and does not extend to the rock; the oil rises within 10 feet of the well, which contains 13,724 gallons, or 343 barrels of oil, has been in operation over two years." Gale also reported that Drake did not commence drilling until June 1859. Presently something of a consensus has been arrived at with the curators of the Drake Well Museum in Titusville, Pennsylvania, acknowledging that Williams *dug* well came first, but holding on to the claim that Drake produced the first *drilled* well. The curators of the Oil Museum of Canada in Oil Springs and the inhabitants of Enniskillen Town-

ship argue that the birthplace of the modern petroleum industry lies in Enniskillen Township and that Williams remains "the father of the modern oil industry." See Harkness/Miller Correspondence, Box 11 [1] A-A, Lambton Room, Lambton County Library, Wyoming, Ontario.

40 *Sarnia Observer*, 30 December 1858.

41 *Hamilton Spectator*, 4 July 1860.

42 Zaslow, *Reading the Rocks*, 15; Harkness, "Makers of Oil History."

43 Harkness, "Makers of Oil History," 137–8; Lucier, "A Plea for Applied Geology," 283–318.

44 Gesner, *A Practical Treatise on Coal Petroleum and Other Distilled Oils*, 159. See also, Russell, "Abraham Gesner," in *Dictionary of Canadian Biography*, Vol.IX, 309–12.

45 Canada, Census of 1861, Enniskillen mss.

46 *Montreal Gazette*, 1 March 1861.

47 Hunt, "Notes on the History of Petroleum or Rock Oil," 1–15.

48 Ibid., 8.

49 Ibid., 9.

50 Galey, "The anticlinal theory of oil and gas accumulation: Its role in the inception of the natural gas and modern oil industries in North America," in Drake and Jordan, *Geologists and Ideas*, 427.

51 Newell, *Technology on the Frontier*, 34–5.

52 *Sarnia Observer*, 30 August 1861; 24 January 1862.

53 *Globe*, 6 September 1861.

54 *London Free Press*, reprinted in the *Sarnia Observer* 14 February 1862. See also the *Globe*, 28 January 1862.

55 *Sarnia Observer*, 30 August 1861; *Globe*, 6 September 1861.

56 *Sarnia Observer*, 16 August 1861.

57 Phelps, "The Canada Oil Association – An Early Business Combination," 31–9.

58 *Sarnia Observer*, 22 January 1863; 30 January 1863.

59 *Sarnia Observer*, 12 June 1863.

60 *Oil Springs Chronicle*, 22 January 1863. Unfortunately, only a few scattered issues of the newspaper are extant.

61 *Sarnia Observer*, 22 January 1863.

62 *Sarnia Observer*, 3 April 1863.

63 *Sarnia Observer*, 12 June 1863; 23 October 1863.

64 *Globe*, 14 April 1865; Nelles, *The Politics of Development*, 20–1, 48.

65 *Sarnia Observer*, 24 November 1865.

66 Lightman, "The Voices of Nature: Popularizing Victorian Science," in Lightman, *Victorian Science in Context*, 189. See also Cooter and Pumfrey, "Separate Spheres and Public Places," 237–67.

67 *Sarnia Observer*, 5 January 1866.

68 Ibid.

69 White, *Geology, Oil Fields and Minerals of Canada West.*

70 Henry White lived in Toronto during the 1860s and 1870s. He is listed in Might & Taylor's *Toronto Directory for 1877* as a geologist and Provincial Land Surveyor, residing at 14 Nassau Street. He published another guide in 1867 entitled, *Gold Regions of Canada. Gold: How and Where to find It! The Explorer's Guide and Manual of Practical and Instructive Directions for Explorers and Miners in the Gold Regions of Canada.* On 3 January 1879 White died in Beaverton, Ontario. See Wallace, *The Macmillan Dictionary of Canadian Biography*, 882; Morgan, *The Dominion Annual Register and Review for 1879*, 427; Morgan, *Bibliotheca canadensis*, 389; Might & Taylor, *Toronto Directory for 1877*, 391; McEvoy, *Toronto Directory for 1868*, 341.

71 White, *Geology, Oil Fields, and Minerals*, 3–4. See Chapman, *A Popular and Practical Exposition of the Minerals and Geology of Canada.*

72 Secord, *Victorian Sensation*, 46–57.

73 White, *Geology, Oil Fields, and Minerals*, 8–9.

74 Ibid., 52–3.

75 Ibid., 69.

76 Ibid., 64–5.

77 Rudwick, "The Emergence of a Visual Language for Geological Science 1760–1840," 151, 159; Harrison, "Nature and Significance of Geological Maps," in Albritton Jr, *The Fabric of Geology*, 225–32.

78 Ibid., 161.

79 Winchell, *Sketches of Creation*, v-vii.

80 Malone, *Dictionary of American Biography, volume 20*, 374–4; *http://www.hti.umich.edu/chi/f/fin...subview=outline&id-umich-bhl-86321*, accessed 11/02/2003. Winchell was named the first chancellor of Syracuse University in 1873. He resigned the following year, but remained at Syracuse as professor of geology until 1876, when he accepted a position as professor of geology and zoology at Vanderbilt University. He was fired in 1878 after a dispute over the teaching of evolution. Shortly thereafter he returned to the University of Michigan, where he remained until his death in 1891.

81 Winchell, *Sketches of Creation*, vii.

82 Secord, *Victorian Sensation*, 56–7.

83 Winchell, *Sketches of Creation*, 13.

84 Ibid., 103.

85 Ibid., 67, 281.

86 Ibid., 273.

87 Ibid., 289–90.

88 Ibid., 282.
89 Ibid., 286–7.
90 Ibid., 287.
91 *The Mail*, 1 December 1888; Winchell, *Sketches of Creation*, 286–7.
92 Owen, *Trek of the Oil Finders*, xiii.
93 Winchell, *Sketches of Creation*, 275.
94 *Globe*, 14 April 1865.
95 Barrett and Besterman, *The Divining-Rod: An Experimental and Psychological Investigation*, xxiv. In the early part of the twentieth-century Barrett, who was a principal founder of the Society for Psychical Research in London, posited a "scientific" explanation for the movement of the rod. In *The Divining-Rod* Barrett and Besterman argued that the movement of the rod was due to the unconscious muscular action of the operator.
96 *Sarnia Observer*, 15 December 1865.
97 *Sarnia Observer*, 23 March 1866.
98 Winchell, *Sketches of Creation*, 273.

CHAPTER THREE

1 Belden, *Belden's Illustrated Historical Atlas of the County of Lambton 1880*, 10.
2 The account was reprinted in the Petrolia *Advertiser* on 12 April 1899.
3 Williams, *The County and the City*, 1, 103–4, 164.
4 Shields, *Places on the Margin*; Soja, *Postmodern Geographies*; Cohen, *Symbolising boundaries*; Thrift and Williams, *Class and space*; Massey, *Space, Place, and Gender.*
5 Burley, *A Particular Condition in Life*; Holman, *A Sense of Their Duty.*
6 Mangan and Walvin, *Manliness and morality*; Carnes and Griffen, *Meanings For Manhood*; Johnson, "Bulls, Bears, and Dancing Boys," 5–37.
7 *Globe* 12 September 1861. James Kerr, builder, and Mayor of Petrolia from 1879, settled in Oil Springs in 1862, and moved to Petrolia in 1869. In a descriptive history of Petrolia, penned by Kerr for the Toronto *Mail* on 1 December 1888, he indicated that oil was struck on lot 13, concession 10. Local historian Victor Lauriston also located the first Petrolia wells on lot 13, concession 10. See Lauriston, *Lambton's Hundred Years 1849–1949*, 145.
8 Reprinted in the *Sarnia Observer and Lambton Advertiser*, 30 August 1861.
9 *Sarnia Observer and Lambton Advertiser*, 31 May 1861.
10 *London Free Press*, 25 June 1861; *Globe*, 12 September 1861, 13 March 1862.
11 *Globe*, 12 September 1861.
12 J.H. Fairbank, "Sketch of Petrolia's Early Days," 3.
13 *Sarnia Observer*, 24 November 1865.
14 *Sarnia Observer*, 30 June 1865.

15 *Globe,* 13 March 1862; *Sarnia Observer,* 13 October 1865; 10 November 1865.
16 *Sarnia Observer,* 29 December 1865.
17 *Sarnia Observer,* 15 December 1865.
18 *Sarnia Observer,* 12 January 1866; 19 January 1866; Phelps, "John Henry Fairbank of Petrolia;" Beers, *Commemorative Biographical Record of the County of Lambton Ontario,* 11.
19 *Advertiser-Topic,* 18 December 1924, 16 July 1980, 12 July 1979; Whipp and Phelps, *Petrolia 1866–1966,* 12–13. When the bank suffered a setback with the development of the Pennsylvania oil territory in 1873, the Bank of Montreal stood behind it. The "little red bank" lasted until 1924. It was one of the last private banks in Canada.
20 *Globe,* 18 January 1867.
21 *Sarnia Observer,* 30 November 1866, 21 December 1866; *Globe,* 18 January 1867.
22 *Sarnia Observer,* 30 November 1866, 21 December 1866.
23 See Griffin, "Petrolia, Cradle of Oil-Drillers," 20.
24 *Advertiser-Topic,* 8 March 1928.
25 *Sarnia Observer,* 2 March 1966.
26 McEvoy, *Gazetteer and Directory of the Counties of Kent, Lambton, and Essex, 1866–7,* 277.
27 Phelps, "John Henry Fairbank," 53; *Sarnia Observer,* 21 December 1866.
28 John D'Oyly Noble, the son of a clergyman, was born on 17 November 1835 in County Meath, Ireland. He apprenticed as a clerk in a linen business in Ballymere before immigrating to New York City. He worked for John Gibon & Co. for two years until 1854 when the business failed. He eventually made his way to Kingston, Ontario, where he re-entered business as the owner of a vessel on the Great Lakes. He married Helen Kirpatrick, the daughter of Stafford Kirpatrick, a prominent Kingston judge, on 26 August 1969. During his voyages on the Great Lakes, Noble became aware of the developments in the Enniskillen oil region and relocated with his family to Petrolia. John Noble brought considerable capital with him, which he invested in oil property. One of the most productive wells in the Petrolia field was "Noble No. One," in the immediate vicinity of the King flowing well. He was also active in local government and was elected to the first town council in 1874. See *Advertiser-Topic,* 9 March 1922, 23 March 1922.
29 *Report of the Royal Commission on the Mineral Resources of Ontario and Measures for Their Development,* 165. The royal commission was struck by the Mowat government on 16 May 1888 to study the province's mineral resources and to recommend the appropriate political and administrative reforms necessary to speed their development. See Nelles, *The Politics of Development,* 25–7.

30 Harris, "Industry and the Good Life around Idaho Peak," 315–43.
31 *Sarnia Observer*, 4 January 1867.
32 *Sarnia Observer*, 2 April 1869.
33 Whipp and Phelps, *Petrolia 1866–1966*, 11.
34 *Advertiser-Topic*, 20 April 1939, 27 July 1939.
35 *Sarnia Observer*, 2 April 1869.
36 *Sarnia Observer*, 30 July 1869.
37 *Sarnia Observer*, 27 May 1870.
38 *Sarnia Observer*, 9 June 1871.
39 *Sarnia Observer*, 4 July 1873; *Advertiser-Topic*, 11 July 1963.
40 Beers, *Commemorative Biographical Record of the County of Lambton Ontario*, 10–11.
41 *Sarnia Observer*, 4 August 1871; Whipp and Phelps, *Petrolia 1866–1966*, 63.
42 *Globe*, 30 August 1881.
43 Ibid.
44 *The Mail*, 1 December 1888. Kerr's article was reprinted as "An Early View of Petrolia, Ontario," in the *Western Ontario Historical Notes*, 17, 2 (September 1962), 57–91.
45 Frager, "Labour History and the Interlocking Hierarchies of Class, Ethnicity, and Gender," 217–47; Parr, *The Gender of Breadwinners*, 123–64.
46 Bond, *Canadian Directories, 1790–1987*, xx. See also Shaw, "Nineteenth Century Directories as Sources in Canadian Social History," 107–21; Davidoff and Hall, *Family Fortunes*; Morris, *Class, sect and party*, 22–3.
47 Crossick, "The Emergence of the Lower Middle Class in Britain," 14.
48 Acheson, "Changing Social Origins of the Canadian Industrial Elite, 1880–1910," 189–216; Acheson, "The Social Origins of the Canadian Industrial Elite, 1880–1885," 144–217; Holman, *A Sense of Their Duty*, 28–49. The occupation as reported in the census was used to determine class designation for statistical purposes in Table 3.6.
49 Phelps, "John Henry Fairbank."
50 Grant, "The 'Mysterious' Jacob L. Englehart and the Early Ontario Petroleum Industry," 66.
51 *The Mail*, 1 December 1888.
52 Newell, *Technology on the Frontier*, 121–34.
53 *Report of the Royal Commission on the Relations of Labor and Capital, 1889*, Ontario Evidence, 695, 710.
54 Ibid., 81–798.
55 Baskerville, "She Has Already Hinted at Board," 105–27.
56 Grant and Thille, "Tariffs, Strategy, and Structure," 390–413.
57 Bliss, *A Living Profit*, 34–5, 45; Ewing, "The History of Imperial Oil Limited."
58 Whipp and Phelps, *Petrolia 1866–1966*, 17; *Sarnia Observer*, 9 June 1871.

59 *Sarnia Observer*, 23 June 1871, 7 July 1871, 16 September 1871; Phelps, "John Henry Fairbank," 73–5.
60 *Sarnia Observer*, 3 May 1872, 17 May 1872.
61 See Harkness, "Makers of Oil History, 1850–1880," for a discussion of the development of the Marthaville field.
62 *Sarnia Observer*, 25 April 1873, 9 May 1873, 16 May 1873, 29 August 1873, 28 November 1873, 25 December 1873.
63 Newell, *Technology on the Frontier*, 121.
64 Phelps, "John Henry Fairbank," 35–6; *Petrolia Advertiser*, 7 June 1878; *Petrolia Topic*, 21 April 1879. The fifteen original stockholders were Messrs Fairbank, Lancey, Vaughn, Cooley, Perkins, Stevenson, Hammond, Lindsay, Gleeson, Harley, Cochrane, McGarvey, Woodward, and Simmons, all of Petrolia, and James M. Williams of Hamilton. The Home Oil Works had its own copper shop. In 1879 it employed 25 men, ten of whom were coopers engaged in the manufacture of new barrels. The labour force was small, but this was attributed to the fact that the distillation work was already carried out at other refineries.
65 *Petrolia Advertiser*, 10 July 1874.
66 *Petrolia Advertiser*, 17 July 1874.
67 *Petrolia Advertiser*, 13 November 1874.
68 *Petrolia Advertiser*, 29 January 1875, 5 February 1875, 19 March 1875.
69 *Petrolia Advertiser*, 7 May 1875, 4 June 1875.
70 *Petrolia Advertiser*, 25 June 1875.
71 *Petrolia Advertiser*, 19 October 1877.
72 *Petrolia Advertiser*, 2 November 1877, 9 November 1877.
73 *The Monetary Times*, 18 April 1879.
74 *Petrolea Topic*, 8 May 1879; *Petrolia Advertiser*, 9 May 1879.
75 *Petrolia Advertiser*, 13 June 1879.
76 *Petrolea Topic*, 9 October 1879.
77 *Petrolia Advertiser*, 31 October 1879.
78 *Monetary Times*, 15 August 1879; *Petrolea Topic*, 13 May 1880.
79 *Petrolea Topic*, 11 December 1879.
80 *Petrolea Topic*, 22 April 1880.
81 Ewing, "The History of Imperial Oil Limited," 59; Grant, "The 'Mysterious' Jacob L. Englehart," 73.
82 *Monetary Times*, 16 July 1880, 6 August 1880, 10 September 1880.
83 *Petrolia Advertiser*, 2 January 1884.
84 *Petrolia Advertiser*, 26 September 1884.
85 The charter members of the Petrolia Oil Exchange were as follows: John H. Fairbank, Jacob L. Englehart, Robert Morris, William K. Gibson, James Kerr,

William H. Hammond, George Sanson, Robert D. Noble, John Walker, Richard I. Bradley, Martin J. Woodward, John D. Noble, Daniel M. Kennedy, Benjamin S. VanTuyl, Edwin D. Kerby, John Brake, James S. Loughead, James McMillan, James McCort, Alexander Clark Edward, Charles Jenkins, Adam Isbister, William McKay, Harrison Corey, Romeyn Lawyer, Joseph A. Grant, Oliver W. Chamberlin, Alfred T. Gurd, John McLister, John Forestal, George Palmer, Frederick Wolfe, James Duffield, and Frederick A. Fitzgerald. See *The Constitution and Rules of the Petrolia Oil Exchange*, Adopted 9 April 1885 (Petrolia 1885).

86 *Constitution and Rules of the Petrolia Oil Exchange.*
87 *Report of the Royal Commission on the Relations of Labor and Capital, 1889*, Ontario Evidence, 717–18.
88 *Petrolia Advertiser*, 26 June 1885.
89 *Petrolia Advertiser*, 31 October 1884.
90 *Report of the Royal Commission on the Relations of Labor and Capital, 1889*, 696.
91 *Petrolia Advertiser*, 21 November 1884, 28 November 1884.
92 *Petrolia Advertiser*, 11 September 1885.
93 *Petrolia Advertiser*, 10 February 1888.
94 *Report of the Royal Commission on the Relations of Labor and Capital, 1889*, 707–10; *Report of the Royal Commission on the Mineral Resources of Ontario and Measures for Their Development*, Toronto 1890, 164.
95 Ibid., 693.
96 Kealey and Palmer, *Dreaming of what might be*, 16–17, 64. For a discussion of the importance of Christianity in the articulation of working-class consciousness in the Knights of Labor in small-town Ontario see Marks, "The Knights of Labor and the Salvation Army," 89–127.
97 *Petrolia Advertiser*, 19 March 1886.
98 *The Palladium of Labor*, 21 August 1886; Kealey and Palmer, *Dreaming of what might be*, 368–9.
99 *Report of the Royal Commission on the Mineral Resources of Ontario*, 160.
100 Ibid., 166.
101 Ewing, "The History of Imperial Oil Canada," 232–3.
102 *Petrolia Advertiser*, 3 June 1897.
103 *Petrolia Advertiser*, 29 June 1898.

CHAPTER FOUR

1 *Petrolia Advertiser*, 11 February 1914.
2 Beers, *Commemorative Biographical Record of the County of Lambton Ontario*, 45.
3 Parr, *The Gender of Breadwinners*, 187–8.

4 Phelps, "John Henry Fairbank," 65.
5 Ibid., 224.
6 *Petrolia Advertiser,* 19 March 1896.
7 Kathleen McCarthy argues that the enduring caricature of "Lady Bountiful" has served to trivialize women's philanthropy. She points out that nineteenth-century American women most often turned to non-profit institutions and reform associations as their primary points of access to public roles, and in the process they forged power structures parallel to those used by men. McCarthy, *Lady Bountiful Revisited,* 1–31.
8 Davidoff and Hall, *Family Fortunes,* 419–20.
9 Clawson, *Constructing Brotherhood, Class, Gender and Paternalism,* 11.
10 *Petrolia Advertiser,* 20 November 1874.
11 *Golden Jubilee, Washington Lodge, No. 260, G.R.D., A.F. & A.M., Petrolea, Canada, 1871–1921.* Lambton Room, Lambton County Library, Wyoming, Ontario.
12 *Petrolia Advertiser,* 16 January 1901.
13 *Advertiser-Topic,* 6 February 1919.
14 *Petrolia Advertiser,* 4 July 1887.
15 *Petrolia Advertiser,* 24 June 1987.
16 Marks, *Revivals and Roller Rinks,* 109.
17 Anstead, "Fraternalism in Victorian Ontario"; Sutherland, "Voluntary Societies and the Process of Middle-class Formation in Early-Victorian Halifax, Nova Scotia," 237–63.
18 Palmer, *Working Class Experience,* 97; Palmer, *A Culture in Conflict,* 39. In a critique of Marks's *Revivals and Roller Rinks,* Palmer reiterates that the "interpretive complexities" of working-class formation must be considered in analysing associational culture. He reiterates the argument that he made in 1979 in *A Culture in Conflict,* that the lessons workers learned within fraternal lodges were varied and complex. See Palmer, "Historiographic Hassles," 139–40.
19 Robert Morris makes this argument in his study of middle-class formation in nineteenth-century Leeds. See Morris, *Class, sect and party,* 167.
20 Newbury, "No Atheist, Eunuch or Woman."
21 Anstead, "Fraternalism in Victorian Ontario;" Morris, *Class, sect and party,* 167–9; Sutherland, "Voluntary Societies and the Process of Middle-class Formation," 237–63.
22 Beers, *Commemorative Biographical Record of the County of Lambton Ontario,* 224–5.
23 Marks, *Revivals and Roller Rinks,* 111; Emery and Emery, *A Young Man's Benefit,* 3–17.
24 J.D. Noble, testimony, *Report of the Royal Commission on the Mineral Resources of Ontario and Measures for their Development,* 165.

25 *Souvenir Program, 46th Annual Convention and Demonstration, Firemen's Association of Ontario,* 1–4 August 1947, 41.
26 *Petrolia Topic,* 8 June 1879.
27 Petrolia Fire Log, 1884–1894, Lambton Room, Lambton County Library, Wyoming, Ontario.
28 Davis, *Parades and Power,* 73–111.
29 *Petrolia Advertiser,* 4 July 1879.
30 *Petrolia Advertiser,* 9 April 1886.
31 *Petrolia Advertiser,* 14 May 1875.
32 *Petrolia Advertiser,* 29 November 1878.
33 *Petrolea Topic,* 24 April 1879, 8 May 1879, 8 August 1879.
34 *Petrolea Topic,* 18 September 1879, 23 September 1879.
35 *Petrolia Advertiser,* 25 June 1880.
36 *Petrolea Topic,* 10 March 1891.
37 *Petrolia Advertiser,* 13 November 1891.
38 *Petrolia Advertiser,* 11 May 1894.
39 *Petrolia Advertiser,* 2 April 1896.
40 Moss, *Manliness and Militarism,* 21–35.
41 O'Brien, "Manhood and the Militia Myth," 125.
42 Marks, *Revivals and Roller Rinks,* 119–20.
43 Moss, *Manliness and Militarism,* 33–4; O'Brien, "Manhood and the Militia Myth," 132–3.
44 *Petrolea Topic,* 18 September 1879.
45 *Petrolia Advertiser,* 19 September 1879; *Topic,* 25 September 1879.
46 Murray, *Come, Bright Improvement,* xi, 5–7, 10.
47 *Petrolea Topic,* 24 April 1879.
48 Murray, *Come, Bright Improvement,* 104–5.
49 *Petrolia Advertiser,* 2 May 1879.
50 *Petrolea Topic,* 24 April 1879; Bruce, *Free Books for All,* xi, 15–16.
51 *Petrolea Topic,* 1 May 1879.
52 *Petrolea Topic,* 19 June 1879.
53 *Petrolea Topic,* 8 May 1879.
54 Morris, *Class, sect and party,* 197.
55 *Petrolea Topic,* 10 June 1880.
56 *Petrolea Topic,* 5 August 1880; *Petrolia Advertiser,* 25 March 1881. For a history of the mechanics' institutes in Ontario and adult education see Vernon, "The Development of Adult Education in Ontario 1790–1900."
57 A handwritten insertion in the Constitution and By-laws of the Petrolia Mechanics' Institute lists the periodicals supplied in the reading room as follows: Toronto *Globe,* Toronto *Mail,* London *Advertiser,* Sarnia *Canadian,*

Sarnia Observer, New York Herald, Chicago *Inter Ocean*, Bradford *Era, Oil City Derrick*, Hamilton *Spectator, Irish Canadian*, Montreal *Daily Witness*, Petrolea *Topic*, Petrolia *Advertiser*, Strathroy *Dispatch*, St Mary's *Argus, Engineering & Mechanics Journal, Journal of Commerce, Dominion Churchman, Scientific American, Monetary Times, Grip, The Bystander, Harpers Weekly, Harpers Monthly, Scribners' Monthly, Canadian Illustrated News, Scientific Canadian, Chambers Journal, Good Words*, and the *Atlantic Monthly*. Charles Fairbank, Petrolia, Ontario, private collection.

58 Martha Corey, a schoolteacher, Mattie Whittaker, a milliner, and John Alfred Webster, a student and son of a farmer, were subscribers to the lending library.

59 This parallels Newbury's findings for the Ottawa Mechanics' Institute and Lynne Marks' findings for Campbellford and Thorold. See Newbury, "No Atheist, Eunuch or Woman"; Marks, *Revivals and Roller Rinks*, 125.

60 *Petrolia Advertiser*, 29 May 1885, 5 June 1885, 27 January 1888, 3 February 1888. See also Vernon, "The Development of Adult Education in Ontario," 154.

61 Charles Howell to the Board of Trustees of the Petrolia Mechanics' Institute, 23 December 1889, Charles Fairbank, Petrolia, Ontario, Private Collection.

62 *Petrolia Advertiser*, 4 March 1887, 18 March 1887, 1 April 1887, 14 April 1887, 13 May 1887, 25 November 1887, 10 December 1887, 23 December 1887, 27 January 1888.

63 *The Mail*, 1 December 1888.

64 *Advertiser-Topic*, 11 December 1924, 18 December 1924. W.A. Fraser was born in Pictou County, Nova Scotia, in 1859. He was educated in Boston, New York, and India, and later worked in India, Burma, and the Canadian Northwest. He settled in Georgetown, Ontario. Fraser established an early reputation as a teller of animal stories. Literary historian Carl Klinck notes that Fraser was one of the more effective magazine story-writers of his day; he entertained his audience with a lively story, sharply realized characters, accurate dialect, and a lack of sentimentality. See Klinck, *Literary History of Canada*, 293, 324–5, 628.

65 *Advertiser-Topic*, 25 December 1924.

66 *Petrolea Topic*, 29 May 1879, 3 July 1879, 17 July 1879, 5 August 1880.

67 Kasson, *Rudeness and Civility*, 215.

68 Ibid., 3.

69 *Petrolia Advertiser*, 4 December 1891, 8 January 1892.

70 *Petrolia Advertiser*, 27 January 1893.

71 Ibid.

72 Davis, *Parades and Power*, 5. See also Goheen, "Symbols in the Streets," 237–43; Heron and Penfold, "The Craftsmen's Spectacle," 358–90; Huskins, "A Tale of Two Cities," 31–46.

73 Bouchier, *For the Love of the Game*, 31–59.
74 *Petrolia Advertiser*, 28 May 1878, 31 May 1878.
75 *Petrolia Advertiser*, 3 July 1885.
76 Heron and Penfold, "The Craftsmen's Spectacle," 371–3.
77 *Petrolia Advertiser*, 3 July 1885.
78 *Petrolia Advertiser*, 8 July 1892.
79 Ryan, *Women in Public*, 19–35; Morgan, *Public Men and Virtuous Women*, 183–214.
80 *Petrolia Advertiser*, 8 July 1892. See also Burr, "Gender, Sexuality, and Nationalism in J.W. Bengough's Verses and Political Cartoons," 505–54, for a discussion of the gender iconography of the Canadian nation.
81 Howell, *Blood, Sweat, and Cheers*, 29; Bouchier, *For the Love of the Game*, 31–59.
82 *Advertiser-Topic*, 29 July 1981; Bouchier, *For the Love of the Game*, 51; Morton, *At Odds*, 8.
83 Howell, *Blood, Sweat, and Cheers*, 28–50; Bouchier, *For the Love of the Game*, 3–7; Burstyn, *The Rites of Men*, 45–75; Wamsley, "The Public Importance of Men and the Importance of Public Men," 24–39; Wise, "Sport and Class Values in Old Ontario and Quebec," 93–117.
84 Burstyn, *The Rites of Men*, 60.
85 Howell, "Baseball, Class and Community in the Maritime Provinces, 1870–1910," 265–86; Palmer, *A Culture in Conflict*, 52–3; Voisey, *Vulcan*, 164–5.
86 *Petrolia Advertiser*, 6 July 1877.
87 *Petrolea Topic*, 27 June 1890.
88 *Petrolia Advertiser*, 13 May 1892, 9 June 1893.
89 *Petrolia Advertiser*, 26 May 1893.
90 *Petrolia Advertiser*, 10 August 1894, 17 August 1894.
91 *Petrolia Advertiser*, 1 August 1895; Gordon, "Any Desired Length," 24–51.

CHAPTER FIVE

1 Lauriston, "The Town of World Travelers," 18. Unfortunately the issue of the *Petrolia Advertiser* containing a description of their departure is no longer available.
2 Pratt, *Imperial Eyes*, 201–27.
3 *Petrolia Advertiser*, 18 October 1874.
4 Ibid.
5 Ibid. Although "a good show of oil" was reported initially by the drilling crew who went to Tjibodas in 1873, the yield proved too small to be commercially viable, and the search for oil was abandoned owing to a lack of capital. Cana-

dian drillers from Enniskillen Township would, however, return to the Dutch East Indies later in the 1890s. See Ooi, *The Petroleum Resources of Indonesia*, 2.

6 Lauriston, "The Town of World Travelers," 19.

7 Kent, *Oil and Empire*; Fischer, *Oil Imperialism*; Ferrier, *The History of the British Petroleum Company, Volume 1*.

8 Burton, "Rules of Thumb," 483–500.

9 McClintock, *Imperial Leather*; Marks, "History, the Nation and Empire," 111–19; Mackenzie, *Imperialism and Popular Culture*; Tabili, *We Ask for British Justice*; Hall, "Rethinking Imperial Histories," 3–29; Burton, "Who Needs the Nation," 228–9.

10 Berger, *The Sense of Power*.

11 Buckner, "Whatever happened to the British Empire," 3–4. Colin Coates suggested that the "limited identities" approach to the writing of Canadian history has done little to address the significance of the majority identities, including the flow of British immigrants to Canada from 1867 to 1914, whose experience was shaped by the Canada-British connection. See Coates, *Imperial Canada, 1967–1916*.

12 Pratt, *Imperial Eyes*; Spurr, *The Rhetoric of Empire*; Hall, "Going-a-Trolloping," 180–99; Dubinsky, *The Second Greatest Disappointment*; Dubinsky, "Local Colour," 67–79; Stoler, "Rethinking Colonial Categories," 134–61.

13 Pratt, *Imperial Eyes*, 4–7.

14 *Petrolia Advertiser*, 18 November 1897.

15 Stowe, *Going Abroad*, 55–73.

16 Pratt, *Imperial Eyes*. I am also indebted to Karen Dubinsky's discussion of travel writing and tourism in Burton, *Gender, Sexuality, and Colonial Modernities*, 67–79.

17 Lauriston, *Lambton's Hundred Years 1849–1949*, 188–9; *Petroleum World* (May 1922), 206; Newell, *Technology on the Frontier*, 128.

18 Pratt and Morningstar, "Early Development of Oil Technology," 3–9; Newell, *Technology on the Frontier*, 34–5; *Petroleum World* (June 1917), 247–8.

19 *Petrolea Topic*, 6 January 1903; Whipp and Phelps, *Petrolia, 1866–1966*, 64; *Belden's Illustrated Historical Atlas of the County of Lambton*, 33.

20 *Petrolia Advertiser*, 19 December 1895. A brief biography of Countess Zeppelin was published in the *Petrolea Topic*, 10 March 1915.

21 *Petrolea Topic*, 27 September 1889.

22 *Report of the Royal Commission on the Mineral Resources of Ontario and Measures for their Development*, 160–1.

23 Most of the information about Gillespie's early years was gleaned from his personal memoirs recorded by his daughter Kathleen Gillespie when her

father was over ninety years of age. My thanks to Kathleen Gillespie for sharing this document with me. See also Lauriston, "Romantic Career of 'Bill' Gillespie," 3–14.

24 Lauriston, "The Town of World Travelers," 19.

25 Whipp and Phelps, *Petrolia 1866–1966*; May, *Hard Oiler*, 114–17.

26 Lauriston, "The Town of World Travelers," 19.

27 Many of the extant travel writings were penned by Enniskillen drillers who went to the East Indies. In 1880 A.J. Zijker, Dutch manager of the East Sumatra Tobacco Company, discovered oil pools in the vicinity of his plantation. He secured a concession to the oil lands, officially named Telaga Said, from the Sultan of Langkat. The discovery of oil in commercial quantities at Telaga Said provided the impetus for exploration and drilling in Sumatra, North Borneo, Brunei, Sarawak, and Java. In 1890 Zijker signed over his petroleum concessions to the newly formed Royal Dutch Company. Some of the drillers who went to the Dutch East Indies in the 1890s and early 1900s signed with the Shell Transport and Trading Company founded by an Englishman, Marcus Samuel. Financial difficulties forced the merger of the Shell Transport and Trading Company with Royal Dutch Shell in 1907. See Ooi, *The Petroleum Resources of Indonesia*, 2–3; *Petrolia Advertiser*, 3 September 1896; *Petroleum World*, 21 October 1905, April 1921.

28 *Petrolia Advertiser*, 18 January 1899, 4 October 1899; *Petrolea Topic*, 13 April 1899.

29 In August 1895 the *Petrolia Advertiser* printed an excerpt from the *Detroit Tribute* describing Townsend's visit to the city. According to the account, Townsend, during a residence of thirteen years in India, had accumulated a considerable fortune and had organized the East India Mining Company with capital of three million pounds sterling. While en route to the Russian oil fields in the fall of 1898, he became ill and was forced to return to London, England, where he died on 10 October. See *Petrolia Advertiser*, 1 August 1895, 19 October 1898.

30 *Petrolia Advertiser*, 3 December 1880, 10 December 1880.

31 *Petrolia Advertiser*, 3 December 1880.

32 Said, *Orientalism*, 3, 41. Said suggested that the Suez Canal destroyed the Orient's distance, "its cloistered intimacy," from the West. According to Said, the Orient was transformed from resistant hostility into obliging and submissive partnership. Townsend's gaze points to Western domination, but there is still evidence of resistance from Arab peoples in his narrative.

33 *Petrolia Advertiser*, 3 December 1880.

34 Spurr, *The Rhetoric of Empire*, 173–5.

35 May, *Hard Oiler*, 177–8.

36 Personal Diary of Edward Winnett, 22 November 1897, Oil Museum of Canada, Oil Springs, Ontario.
37 *Petrolia Advertiser*, 10 December 1880.
38 *Petrolia Advertiser*, 26 October 1898. In a similar vein M.J. Kelly wrote that Port Said was the "hardest place in the world," with the chief occupation being begging and peddling. See *Petrolia Advertiser*, 11 April 1900.
39 *Petrolia Advertiser*, 10 December 1880.
40 *Petrolia Advertiser*, 26 October 1898.
41 Winnett diary, 23 November 1897; Marks, *Revivals and Roller Rinks*, 32–3.
42 *Petrolia Advertiser*, 10 December 1880.
43 The "wild North West" of Canada was also the scene of adventure stories published in *The Boy's Own Paper*. The heroes were nearly always of British background. See Moyles and Owram, *Imperial Dreams and Colonial Realities*; Bratton, "Of England, Home and Duty," 73–93; Reid and Washbrook, "Kipling, Kim and Imperialism," 14–20.
44 *Petrolia Advertiser*, 1 March 1899.
45 *Petrolia Advertiser*, 1 August 1900.
46 *Petrolia Advertiser*, 30 July 1880.
47 *Petrolia Advertiser-Topic*, 20 June 1922.
48 Pratt, *Imperial Eyes*, 15–35; 69–140; Spurr, *The Rhetoric of Empire*; Said, *Orientalism*, 5–7, 49–110.
49 Contract between William Oliver Gillespie and the Nederlandsch Indische Industrie en Handel Maatschappu, 22 November 1904, Oil Museum of Canada, Oil Springs, Ontario.
50 Agreement between The First Exploitation Company Ltd, England, and Ernest Edward Kells, 1 March 1928, Oil Museum of Canada, Oil Springs, Ontario.
51 Contract between William Oliver Gillespie and the Nederlandsch Indische Industrie en Handel Maatschappu, 22 November 1904; Contract between William Oliver Gillespie and The Anglo-Saxon Petroleum Co. Ltd, 12 July 1920, Oil Museum of Canada, Oil Springs, Ontario.
52 Winnett diary, 16–17 December 1897.
53 *Petrolia Advertiser*, 1 March 1899.
54 Winnett diary, 12 September 1898.
55 See *Petrolia Advertiser*, 1 November 1899, 31 May 1899; *Advertiser-Topic*, 22 July 1920, 19 October 1922.
56 *Advertiser-Topic*, 19 November 1925.
57 *Advertiser-Topic*, 27 December 1923.
58 My thinking here has been influenced by my reading of David Roediger's *The Wages of Whiteness: Race and the Making of the American Working Class*.

59 *Advertiser-Topic,* 12 July 1922.

60 *Petrolea Topic,* 1 July 1880.

61 Ibid.

62 *Petrolia Advertiser,* 11 April 1900; Robert Reade, "Memories of Malaya," *Toronto Star Weekly,* 30 January 1932.

63 Stoler, "Carnal Knowledge and Imperial Power," 221; Stoler, *Capitalism and Confrontation in Sumatra's Plantation Belt, 1870–1979,* 31–5.

64 *Petrolea Topic,* 22 November 1900.

65 "The late John W. Crosbie," *Petrolia Advertiser,* 30 August 1899.

66 Reade, "Memories of Malaya."

67 Lauriston, "Bill Gillespie Comes Home," *The London Advertiser,* 28 August 1926.

68 Copy of Memorandum, Sarawak Oilfields Limited, Miri, 11 July 1923. My thanks to Kathleen Gillespie of Petrolia, Ontario, for providing me with this document.

69 Stoler, "Carnal Knowledge and Imperial Power," 239–44.

70 *Advertiser-Topic,* 16 March 1922, 27 April 1922, 30 November 1922, 22 October 1925.

71 *Advertiser-Topic,* 15 September 1921.

72 *Advertiser-Topic,* 22 February 1923.

73 Vola Braybrook, taped interview with Charles Whipp, 8 June 1984. Copies of the tapes are available at the Petrolia Discovery Museum, Petrolia, Ontario.

74 Agnes Miller, taped interview with Charles Whipp, 5 June 1984.

75 Braybrook interview.

76 Dorothy Stevenson, taped interview with Charles Whipp, 5 June 1984.

77 *Advertiser-Topic,* 25 October 1923.

78 *Advertiser-Topic,* 29 May 1924, 24 November 1927.

79 Agnes Miller, taped interview with Charles Whipp, 5 June 1984.

80 *Advertiser-Topic,* 2 November 1922.

81 Vola Braybrook, taped interview with Charles Whipp, 8 June 1984.

82 Arnold Thompson, taped interview with Charles Whipp, 16 April 1984.

83 Stevenson interview.

84 Ev Brimmer, taped interview with Charles Whipp, 21 May 1985.

85 Stevenson interview.

86 McClintock, *Imperial Leather,* 15.

87 Brimmer interview.

88 Ibid.

89 Stevenson interview.

90 *Petrolia Advertiser,* 17 May 1899; *Advertiser-Topic,* 18 January 1924.

91 *Advertiser-Topic*, 20 July 1922, 15 June 1922, 30 November 1922, 29 October 1925, 8 November 1928.

92 *Advertiser-Topic*, 3 December 1953.

93 Stevenson interview.

94 On 26 September 1890, the *Petrolea Topic* reported: "The Oil Well Supply Co, of Petrolea, are this week shipping a large consignment of drilling tools, supplies, etc., to East India. This firm secured large contracts in those foreign countries, and have [sic] been working night and day for some weeks in order to fill orders."

95 *Petrolia Advertiser*, 24 January 1900.

96 *Petrolea Topic*, 9 August 1900.

CHAPTER SIX

1 Tallentire, "Strategies of Memory," 197; Walsh and High, "Rethinking the Concept of Community," 255–73.

2 Anderson, *Imagined Communities*, 6–15. Alan Gordon suggests that public memory can be distinguished from the broader term "collective memory," by its restriction to a public sphere. This idea of "public memory" has been adopted for the purposes of the discussion here. See Gordon, *Making Public Pasts*, 6.

3 Tallentire, "Strategies of Memory," 198.

4 Samuel, *Theatres of Memory*, ix-17.

5 Lowenthal, *The Past is a Foreign Country*, xvi-28.

6 Halbwachs, *The Collective Memory*, 22–49; Thelen, "Memory and American History," 1122–3.

7 Hayden, *The Power of Place*, 2–45.

8 Samuel, *Theatres of Memory*, 274.

9 Lavenda, "Festivals and the Creation of Public Culture," 100.

10 *Petrolia Advertiser*, 2 October 1907, 9 October 1907.

11 *Petrolia Advertiser*, 17 June 1908.

12 *Petrolia Advertiser*, 16 October 1907.

13 *Petrolea Topic*, 12 August 1908.

14 Ibid.

15 Ryan, *Women in Public*, 19–56; Morgan, *Public Men and Virtuous Women*, 183–218.

16 *Forest Free Press*, 20 August 1908. Unfortunately, there are no extant issues of either the *Petrolea Topic* or the *Petrolia Advertiser* for late August 1908.

17 *Petrolea Topic*, 12 August 1908.

18 *Advertiser-Topic*, 21 August 1924.

19 *Advertiser-Topic*, 4 September 1924.
20 *Advertiser-Topic*, 11 September 1924.
21 *Advertiser-Topic*, 13 November 1924.
22 Boutilier and Prentice, *Creating Historical Memory*, 16; *Advertiser-Topic*, 16 April 1925, 7 May 1925.
23 *Advertiser-Topic*, 25 June 1925. The references to specific remembrances are taken from this issue.
24 *Advertiser-Topic*, 11 June 1925, 18 June 1925, 25 June 1925.
25 The literature pertaining to the commemoration of World War I is now quite extensive. See Olick and Robbins, "Social Memory Studies," 118–19, for a review of the writers who noted a changed in the form of memory after the World War.
26 *Advertiser-Topic*, 9 July 1925.
27 *Advertiser-Topic*, 22 November 1945, 29 November 1945.
28 *London Free Press*, 24 August 1946.
29 Petrolia Old Boys' Reunion, Souvenir Program, 26 August to 2 September 1946, Lambton Room, Lambton County Library, Wyoming, Ontario.
30 *Advertiser-Topic*, 22 August 1946.
31 *Advertiser-Topic*, 29 August 1946.
32 *Advertiser-Topic*, 5 September 1946.
33 *Advertiser-Topic*, 8 September 1946.
34 *Advertiser-Topic*, 22 August 1946.
35 *Advertiser-Topic*, 8 September 1946. Some 70 feet of oil was found standing at the bottom of the hole. A pump was installed. The committee in charge of drilling and shooting the well turned the operation of the well over to the Lambton Central Exhibition Association for the cost of the drilling materials, amounting to $433.60. Two weeks later the well was abandoned and plugged. The well was producing only about two and one-half barrels of oil per day, with the possibility that production would drop. See *Advertiser-Topic*, 19 September 1946.
36 *Advertiser-Topic*, 19 September 1946.
37 *Advertiser-Topic*, 6 August 1946, 22 August 1946.
38 Dicks, *Heritage, Place and Community*, 148–50.
39 *Advertiser-Topic*, 14 February 1973, 2 May 1973.
40 *Advertiser-Topic*, 8 August 1973.
41 Dicks, *Heritage, Place and Community*, 53–4; *Advertiser-Topic*, 12 December 1973, 3 April 1974.
42 *Advertiser-Topic*, 25 September 1974, 2 October 1974, 6 November 1974.
43 *Advertiser-Topic*, 23 June 1976.
44 Dicks, *Culture on Display*, 8–13; McKay, *The Quest of the Folk*, 274–311.

45 *Advertiser-Topic,* 25 July 1979.
46 *Advertiser-Topic,* 27 June 1979, 5 September 1979.
47 *Advertiser-Topic,* 2 July 1980.
48 *Sarnia Observer,* 28 April 1980.
49 Samuel, *Theatre of Memory,* 172–3; Lowenthal, *The Past is a Foreign Country,* 62–3.
50 *Advertiser-Topic,* 14 May 1980.
51 Dicks, *Heritage, Place and Community,* 49.
52 *Advertiser-Topic,* 25 June 1980.
53 *Advertiser-Topic,* 25 June 1980.
54 *Sarnia Observer,* 28 October 1980.
55 *Sarnia Observer,* 7 July 1981.
56 *Sarnia Observer,* 17 June 1981.
57 Dicks, *Culture on Display,* 92–4.
58 The quotation is borrowed from the sign outside Watson's Mill on the museum site.
59 *Advertiser-Topic,* 18 September 1995.
60 Mills, *Crude Beginnings,* 21; *Advertiser-Topic,* 10 June 1998.
61 Mills, *Crude Beginnings,* 22.
62 *Sarnia Observer,* 3 August 1988, 10 August 1988.
63 *Sarnia Observer,* 4 August 1982.
64 *Sarnia Observer,* 10 May 1994, 21 May 1994; *Advertiser-Topic,* 24 May 1995.
65 *Sarnia Observer,* 28 July 1990.
66 *http://www.petroliadiscovery.com,* accessed 16 June 2005.
67 MacRae, *Cornerstones of Order,* 234–9; Fairfield, "Theatres and Performance Halls," 250–2; *Advertiser-Topic,* 19 October 1977, 19 June 1996, 14 September 1992.
68 Brown, "Entertainers of the Road," 123–65.
69 *Advertiser-Topic,* 6 September 1972, 29 October 1972.
70 *Advertiser-Topic,* 21 February 1973; Walsh, *The Representation of the Past,* 72.
71 Survey and Appraisal of buildings of architectural and historic merit in Petrolia. Prepared by The Professional Advisory Committee Architectural Conservancy of Ontario, June 1973. Lambton Room, Lambton County Library, Wyoming, Ontario.
72 *Sarnia Observer,* 18 May 1973.
73 *Advertiser-Topic,* 18 July 1973, 1 August 1973.
74 *Advertiser-Topic,* 25 July 1973, 19 September 1973.
75 Miller, *The Donnellys must die.*
76 *Advertiser-Topic,* 21 November 1973.
77 *Advertiser-Topic,* 19 June 1974, 26 June 1974, 17 July 1974.

78 *Advertiser-Topic*, 5 November 1975.
79 *Advertiser-Topic*, 26 March 1981.
80 *Advertiser-Topic*, 20 September 1982, 18 June 1983; *Sarnia Observer*, 24 June 1983; *London Free Press*, 8 July 1983.
81 *Advertiser-Topic*, 27 May 1981, 17 June 1981.
82 *Sarnia Observer*, 25 January 1989, 26 January 1989.
83 *Sarnia Observer*, 27 January 1989.
84 *Sarnia Observer*, 30 March 1989.
85 *Sarnia Observer*, 13 June 1989, 16 June 1989.
86 *Sarnia Observer*, 19 April 1989, 13 June 1989, 16 June 1989, 11 July 1989, 16 August 1989, 15 September 1989, 21 September 1989, 7 November 1989, 15 November 1989, 28 November 1989.
87 *Sarnia Observer*, 19 December 1989.
88 *Sarnia Observer*, 27 February 1990, 2 March 1990.
89 *Advertiser-Topic*, Victoria Hall 1992 Souvenir Edition, 16 September 1992.
90 *Sarnia Observer*, 20 November 1990, 27 December 1990, 4 January 1991, 11 July 1991, 30 August 1991, 24 September 1991, 1 November 1991, 16 November 1991, 20 December 1991; *Advertiser-Topic*, 23 September 1992.
91 *Sarnia Observer*, 28 September 1992; *Advertiser-Topic*, 16 September 1992, 22 September 1992.
92 *Advertiser-Topic*, 5 June 1996, http://www.petroliaheritage.com accessed 11/06/02.
93 *Advertiser-Topic*, 9 July 1997.

CONCLUSION

1 *Sarnia Observer*, 10 January 1963.
2 *Sarnia Observer*, 29 July 1983.
3 *Petrolea Topic*, 7 August 2002.
4 *London Free Press*, 31 August 2002.
5 *Petrolea Topic*, 18 August 2004, 25 August 2004, 1 September 2004; *London Free Press*, 11 August 2004; *Sarnia Observer*, 10 August 2004, 11 August 2004.
6 Guildford and Morton, *Separate Spheres*, 20–1.
7 National Film Board of Canada, "Hard Oil!" 1991.

Bibliography

Acheson, T.W. "Changing Social Origins of the Canadian Industrial Elite, 1880–1910." *Business History Review* 47, 2 (Summer 1973), 189–216

– "The Social Origins of the Canadian Industrial Elite, 1880–1885." In David S. Macmillan, ed. *Canadian Business History: Selected Studies, 1497–1971*. Toronto: McClelland & Stewart 1972

Alcock, F.J. *A Century in the History of the Geological Survey of Canada*. Ottawa: Department of Mines and Resources 1947

Anderson, Benedict. *Imagined Communities: Reflections on the Origin and Spread of Nationalism*. London: Verso 1991, originally 1983

Anstead, Christopher J. "Fraternalism in Victorian Ontario: Secret Societies and Cultural Hegemony." PhD dissertation, University of Western Ontario, 1992

Armstrong, F.H., et al., eds. *Aspects of Nineteenth-Century Ontario: Essays Presented to James J. Talman*. Toronto: University of Toronto Press 1974

Barrett, William, and Theodore Besterman. *The Divining-Rod: An Experimental and Psychological Investigation*. Foreword by Leslie Shepard. New York: University Books 1906

Baskerville, Peter. "'She Has Already Hinted at Board': Enterprising Urban Women in British Columbia, 1862–1896." *Histoire sociale/Social History* 26, 52 (November 1993), 105–27

Beers & Co. *Commemorative Biographical Record of the County of Lambton Ontario*. Toronto 1906

Belden, H., and Company. *Belden's Illustrated Historical Atlas of the County of Lambton, 1880s*. Revised and published with additions by Edward Phelps. Sarnia 1973

Berger, Carl. *The Sense of Power: Studies in the Ideas of Canadian Imperialism 1867–1914*. Toronto: University of Toronto Press 1970

Bilak, Daniel A. "The Law of the Land: Rural Debt and Private Land Transfer in Upper Canada, 1841–1867." *Histoire sociale/Social History* 20, 39 (May 1987), 177–88

Bitterman, Rusty. "The Hierarchy of the Soil: Land and Labour in a 19th Century Cape Breton Community." *Acadiensis* 17, 1 (Autumn 1988), 33–55

Bliss, Michael. *A Living Profit: Studies in the Social History of Canadian Business, 1883–1911*. Toronto: McClelland & Stewart 1974

Bond, Mary E. *Canadian Directories, 1790–1987: A Bibliography and Place-Name Index*. Ottawa: Canadian Government Publishing Centre 1989

Bouchier, Nancy B. *For the Love of the Game: Amateur Sport in Small-Town Ontario, 1838–1895*. Montreal and Kingston: McGill-Queen's University Press 2003

Boutilier, Beverly, and Alison Prentice, eds. *Creating Historical Memory: English-Canadian Women and the Work of History*. Vancouver: University of British Columbia Press 1997

Bratton, J.S. "Of England, Home and Duty: The Image of England in Victorian and Edwardian Juvenile Fiction." In John M. Mackenzie, ed. *Imperialism and Popular Culture*. Manchester: Manchester University Press 1986

Bray, Matt, and Ashley Thomson, eds. *At the End of the Shift: Mines and Single-Industry Towns in Northern Ontario*. Toronto and Oxford: Dundurn Press 1992

Brown, Mary. "Entertainers of the Road." In Ann Saddlemeyer, ed. *Early Stages: Theatre in Ontario 1800–1914*. Toronto: University of Toronto Press 1990.

Bruce, Lorne. *Free Books for All: The Public Library Movement in Ontario, 1850–1930*. Toronto and Oxford: Dundurn Press 1994

Brunger, A.G. "A spatial analysis of individual settlement in Southern London District, Upper Canada, 1800–1836." PhD dissertation, University of Western Ontario 1973

Buckner, Phillip. "Whatever happened to the British Empire?" *Journal of the Canadian Historical Association*, New Series, Vol. 4 (1993), 3–32

Burley, David G. *A Particular Condition in Life: Self-Employment and Social Mobility in Mid-Victorian Brantford, Ontario*. Montreal and Kingston: McGill-Queen's University Press 1994

Burr, Christina. "Gender, Sexuality, and Nationalism in J.W. Bengough's Verses and Political Cartoons." *Canadian Historical Review*, 83, 4 (December 2002), 505–54

Burstyn, Varda. *The Rites of Men: Manhood, Politics, and the Culture of Sport.* Toronto: University of Toronto Press 1999

Burton, Antoinette, ed. *Gender, Sexuality and Colonial Modernities.* London and New York: Routledge 1999

– "Rules of Thumb: British history and 'imperial culture' in nineteenth- and twentieth-century Britain." *Women's History Review,* 3 (1994), 483–500

– "Who Needs the Nation? Interrogating 'British' History," *Journal of Historical Sociology* 10 (September 1997), 228–9

Carnes, Mark C., and Clyde Griffen, eds. *Meanings For Manhood: Constructions of Masculinity in Victorian America.* Chicago and London: University of Chicago Press 1990

Certeau, Michel de. *The Practice of Everyday Life.* Translated by Steven Rendall. Berkeley: University of California Press 1984

Chambers, Lori. *Married Women and Property Law in Victorian Ontario.* Toronto: University of Toronto Press 1997

Chapman, Edward John. *A Popular and Practical Exposition of the Minerals and Geology of Canada.* Toronto: The Copp, Clark Company 1864

Clarke, John. "Land and Law in Essex County. Malden Township and the Abstract Index to Deeds." *Histoire sociale/Social History* 11, 22 (November 1978), 475–93

– "The Role of Political Position and Family and Economic Linkage in Land Speculation in the Western District of Upper Canada, 1788–1815." *Canadian Geographer* 19, 1 (1975), 18–34

– *Land, Power, and Economics on the Frontier of Upper Canada.* Montreal and Kingston: McGill-Queen's University Press 2001

– and G.F. Finnegan, "Colonial survey records and the vegetation of Essex County, Ontario," *Journal of Historical Geography* 10, 2 (1984), 199–238

Clawson, Mary Ann. *Constructing Brotherhood, Class, Gender and Paternalism.* Princeton, N.J.: Princeton University Press 1989

Clayton, Daniel W. *Islands of Truth: The Imperial Fashioning of Vancouver Island.* Vancouver: University of British Columbia Press 2000

– "Absence, Memory, and Geography." *BC Studies* 132 (Winter 2001/02), 65–79

Coates, Colin M., ed. *Imperial Canada, 1867–1916.* Edinburgh: University of Edinburgh 1997

Cohen, Anthony P., ed. *Symbolising boundaries: Identity and diversity in British cultures.* Manchester: Manchester University Press 1986

Cooper, Frederick, and Ann Laura Stoler. *Tensions of Empire: Colonial Cultures in a Bourgeois World.* Berkeley: University of California Press 1997

Cooter, Roger, and Stephen Pumfrey. "Separate Spheres and Public Places: Reflections on the History of Science Popularization and Science in Popular Culture." *History of Science* 32 (1994), 237–67

Crossick, Geoffrey. *The Lower Middle Class in Britain 1870–1914*. London: Croom Helm 1977

– "The Emergence of the Lower Middle Class in Britain." In Geoffrey Crossick. *The Lower Middle Class in Britain 1870–1914*. London: Croom Helm 1977

Cruikshank, E.A., ed. *The Correspondence of Lieut. Governor John Graves Simcoe, with Allied Documents Related to the Administration of the Government of Upper Canada*. Toronto: The Society 1923

Curtis, Bruce, "Official Documentary Systems and Colonial Government: From Imperial Sovereignty to Colonial Autonomy in the Canadas, 1841–1867." *Journal of Historical Sociology* 10, 4 (December 1997), 389–417

Davidoff, Leonore, and Catherine Hall. *Family Fortunes: Men and Women of the English Middle Class, 1780–1850*. Chicago: University of Chicago Press 1987

Davis, Susan G. *Parades and Power: Street Theatre in Nineteenth-Century Philadelphia*. Philadelphia: Temple University Press 1986

Dicks, Bella. *Culture on Display: The Production of Contemporary Visitability*. London: Open University Press 2003

– *Heritage, Place and Community*. Cardiff: University of Wales Press 2000

Drake, Ellen T., and William M. Jordan, eds. *Geologists and Ideas: A History of North American Geology, Volume I*. Boulder, Colorado: Geological Society of America 1985

Dubinsky, Karen. "Local Colour: The spectacle of race at Niagara Falls." In Antoinette Burton, ed. *Gender, Sexuality and Colonial Modernities*. London and New York: Routledge 1999

– *The Second Greatest Disappointment: Honeymooning and Tourism at Niagara Falls*. Toronto and New Brunswick, New Jersey: Between the Lines 1999

Eagan, William E. "The Canadian Geological Survey: Hinterland Between Two Metropolises." *Earth Sciences History* 12, 2 (1993), 99–106

Edgar, J.D. *A Manual for Oil Men and Dealers in Land*. Toronto: Rollo & Adam 1866

Edney, Matthew H. *Mapping an Empire: The Geographical Construction of British India, 1765–43*. Chicago and London: University of Chicago Press 1997

Elford, Jean Turnbull. *Canada West's Last Frontier: A History of Lambton*. Sarnia: Lambton County Historical Society 1982

Emery, George, and J.C. Herbert Emery. *A Young Man's Benefit: The Independent Order of Odd Fellows and Sickness Insurance in the United States and*

Canada, 1860–1929. Montreal and Kingston: McGill-Queen's University Press 1999

Ewing, John S. "The History of Imperial Oil Limited." Unpublished study for the Business History Foundation Inc., Harvard Business School, 1953, Imperial Oil Archives, Toronto

Fairbank, J.H. "Sketch of Petrolia's Early Days." In *Petrolia Canada, 1862–1908: Its Advantages as a Commercial Residential Centre*. Petrolia: Topic 1908

Fairfield, Robert. "Theatres and Performance Halls." In Ann Saddlemyer, ed. *Early Stages: Theatre in Ontario 1800–1914*. Toronto: University of Toronto Press 1990.

Ferrier, R.W. *The History of the British Petroleum Company, Volume 1: The Developing Years 1901–1932*. Cambridge: Cambridge University Press 1982

Firth, Edith, ed. *Profiles of a Province: Studies in the History of Ontario*. Toronto: Ontario Historical Society 1967

Fischer, Louis. *Oil Imperialism: The International Struggle for Petroleum*. Westport, Connecticut: Hyperion Press 1926

Frager, Ruth A. "Labour History and the Interlocking Hierarchies of Class, Ethnicity, and Gender: A Canadian Perspective." *International Review of Social History* 44 (1999), 217–47

Gagan, David. "The Security of the Land: Mortgaging in Toronto Gore Township, 1834–95." In F.H. Armstrong, et al., eds. *Aspects of Nineteenth-Century Ontario: Essays Presented to James J. Talman*. Toronto: University of Toronto Press, 1974

– "'The Prose of Life': Literary Reflections of the Family, Individual Experience and Social Structure in Nineteenth-Century Canada." *Journal of Social History* 9, 3 (Spring 1976), 367–81

– "Property and Interest: Some Preliminary Evidence of Land Speculation by the Family Compact in Upper Canada, 1820–1840." *Ontario History* 70, 1 (1978), 63–70

Gale, Thomas A. *The Wonder of the Nineteenth Century! Rock Oil, In Pennsylvania and Elsewhere*. Np: Erie & Griffith Publishers 1860

Gates, Lillian F. *Land Policies of Upper Canada*. Toronto: University of Toronto Press 1968

Gentilcore, Louis and Kate Donkin. *Land Surveys of Southern Ontario: An Introduction and Index to the Field Notebooks of the Ontario Land Surveyors*. Toronto: York University 1973

Gesner, Abraham. *A Practical Treatise on Coal Petroleum and Other Distilled Oils*, second edition. New York: A.M. Kelley 1865

Getty, A.L., and Antoine S. Lussier, eds. *As Long as the Sun Shines and Water Flows*. Vancouver: University of British Columbia Press 1983

Goheen, Peter G. "Symbols in the Streets: Parades in Victorian Urban Canada." *Urban History Review/Revue d'histoire urbaine*, 18, 3 (February 1990), 237–43

Gordon, Alan. *Making Public Pasts: The Contested Terrain of Montréal's Public Memories, 1891–1930*. Montreal and Kingston: McGill-Queen's University Press 2001

Gordon, Sarah A. "Any Desired Length: Negotiating Gender Through Sports Clothing, 1870–1925." In Philip Scanton, ed. *Beauty and Business: Commerce, Gender, and Culture in Modern America*. New York: Routledge 2001

Grant, Hugh M. "The 'Mysterious' Jacob L. Englehart and the Early Ontario Petroleum Industry." *Ontario History* 85, 1 (March 1993), 65–76

– and Henry Thille. "Tariffs, Strategy, and Structure: Competition and Collusion in the Ontario Petroleum Industry, 1870–1880." *Journal of Economic History* 61, 2 (June 2001), 390–413

Gregory, Derek. "Imaginative geographies." *Progress in Human Geography* 19, 4 (1995), 447–85

Griffin, Selwyn P. "Petrolia, Cradle of Oil-Drillers," *Imperial Oil Review* 14 (August-September 1930), 20

Guildford, Janet, and Suzanne Morton, eds. *Separate Spheres: Women's Worlds in the 19th Century Maritimes*. Fredericton: Acadiensis Press 1994

Hagopian, John S. "The Use of Land Registry Offices for Historical Research." *Ontario History* 1 (Spring 1995), 77–86

Halbwachs, Maurice. *The Collective Memory*. Translated by Francis J. Ditter, Jr and Vida Yazdi Ditter. New York: Harper & Row 1980

Hall, Catherine. "Rethinking Imperial Histories: The Reform Act of 1867," *New Left Review*, 208 (November/December 1994), 3–29

– ed. *Cultures of Empire: Colonizers in Britain and the Empire in the Nineteenth and Twentieth Centuries*. New York: Routledge 2000

– "Going-a-Trolloping: Imperial Man Travels the Empire." In Clare Midgely, ed. *Gender and Imperialism*. Manchester and New York: Manchester University Press 1998

Harkness, Robert B. "Makers of Oil History, 1850–1880." Unpublished ms, microfilm copy, Regional History Room, University of Western Ontario, nd., available Leddy Library University of Windsor

Harrington, B.J. "Sir William Logan." *American Journal of Science and Arts* 11 (February 1876), 1–13

– *Life of Sir William E. Logan Kt*. London: Sampson Low 1883

Harris, Cole. "Industry and the Good Life around Idaho Peak." *Canadian Historical Review* 66, 3 (1985), 315–43

– *The Resettlement of British Columbia: Essays on Colonialism and Geographical Change.* Vancouver: University of British Columbia Press 1997. Reprinted 2000

Harrison, J.M. "Nature and Significance of Geological Maps." In Claude C. Albritton, Jr, ed. *The Fabric of Geology.* Stanford, California: Freeman, Cooper & Co. 1963

Harvey, David. *The Condition of Postmodernity: An Enquiry into the Origins of Cultural Change.* Oxford and New York: Blackwell, 1989

Hayden, Delores. *The Power of Place: Urban Landscape as Public History.* Cambridge and London: MIT Press 1995

Heick, W.H., and Roger Graham. *His Own Man: Essays in Honour of Arthur Reginald Marsden Lower.* Montreal: McGill-Queen's University Press 1974

Henry, J.T. *The Early and Later History of Petroleum with Authentic Facts in Regard to its Development in Western Pennsylvania.* Philadelphia: Burt Franklin 1873

Heron, Craig. *Working in Steel: The Early Years in Canada, 1883–1935.* Toronto: McClelland & Stewart 1988

– and Steve Penfold. "The Craftsmen's Spectacle: Labour Day Parades in Canada, the Early Years." *Histoire sociale/Social History* 39, 58 (November 1996), 358–90

Hobsbawm, Eric, and Terence Ranger. *The Invention of Tradition.* Cambridge: Cambridge University Press 1983

Holman, Andrew C. *A Sense of Their Duty: Middle-Class Formation in Victorian Ontario Towns.* Montreal and Kingston: McGill-Queen's University Press 2000

Howell, Colin D. "Baseball, Class and Community in the Maritime Provinces, 1870–1910." *Histoire sociale/Social History* 22, 44 (November 1989), 265–86

– *Blood, Sweat, and Cheers: Sport and the Making of Modern Canada.* Toronto: University of Toronto Press 2001

Howison, John. *Sketches of Upper Canada.* Edinburgh: Oliver & Boyd 1821

Hunt, Bruce, J. "Doing Science in a Global Empire: Cable Telegraphy and Electrical Physics in Victorian Britain." In Bernard Lightman, ed. *Victorian Science in Context.* Chicago: University of Chicago 1997

Hunt, T. Sterry. *Notes on the History of Petrolium.* Np. 1861. CIHM no. 63410

Huskins, Bonnie. "'A Tale of Two Cities': Boosterism and the Imagination of Community during the Visit of the Prince of Wales to Saint John and Halifax in 1860." *Urban History Review/Revue d'histoire urbaine* 28, 1 (October 1999), 31–46

Johnson, J.K. *Becoming Prominent: Regional Leadership in Upper Canada, 1791–1841.* Kingston and Montreal: McGill-Queen's University Press 1989

Johnson, Leo. "Land Policy, Population Growth, and Social Structure in the Home District, 1793–1851." *Ontario History* 63 (1971), 41–61

Johnson, Susan Lee. "Bulls, Bears, and Dancing Boys: Race, Gender, and Leisure in the California Gold Rush." *Radical History Review* 60, 4 (1994), 5–37

Jordanova, L.J., and Roy S. Porter, eds. *Images of the Earth: Essays in the History of the Environmental Sciences.* Halfpenny Furze, England: British Society for the History of Science 1979

Karp, Ivan, Christine Mull Kreamer, and Steven D. Lavine. *Museums and Communities: The Politics of Public Culture.* Washington and London: Smithsonian Institution Press 1992

Kasson, John F. *Rudeness and Civility: Manners in Nineteenth-Century Urban America.* New York: Noonday Press 1990

Kealey, Gregory S. "Orangemen and the Corporation: The politics of class during the Union of the Canadas." In Victor L. Russell, ed. *Forging a Consensus: Historical Essays on Toronto.* Toronto: University of Toronto Press 1984

– and Bryan D. Palmer. *Dreaming of what might be: The Knights of Labor in Ontario, 1880–1900.* Toronto: New Hogtown Press 1987. Originally 1982

Kelly, Kenneth. "The Artificial Drainage of Land in Nineteenth-Century Southern Ontario." *The Canadian Geographer* 14, 4 (Winter 1975), 279–98

Kent, Marian. *Oil and Empire: British Policy and Mesopotamian Oil 1900–1920.* New York: Barnes & Noble 1978

Kinda, Akihiro. "The concept of 'township' in Britain and the British colonies in the seventeenth and eighteenth centuries." *Journal of Historical Geography* 27, 2 (2001), 137–52

Klinck, Carl F. *Literary History of Canada: Canadian Literature in English.* Toronto: University of Toronto Press 1965. Reprinted 1973

Knell, Simon J. *The Culture of English Geology, 1815–1851: A science revealed through its collecting.* Aldershot: Ashgate Publishing Co. 2000

Ladell, John L. *They Left Their Mark: Surveyors and Their Role in the Settlement of Ontario.* Toronto and Oxford: Dundurn Press 1993

Lauriston, Victor. *Lambton's Hundred Years 1849–1949.* Sarnia: Haines Frontier Press 1949

– "The Town of World Travelers," *Maclean's Magazine.* 1 May 1924, 18–19

– Romantic Career of 'Bill Gillespie'." *The Rig and Reel Magazine* (October 1927) 3–14

Lavenda, Robert H. "Festivals and the Creation of Public Culture: Whose Voice(s)?" In Ivan Karp et al., eds. *Museums and Communities: The Politics of Public Culture.* Washington: Smithsonian Institution Press 1992

Lefebvre, Henry. *The Production of Space.* Translated by Donald Nicholson-Smith. Oxford, U.K., and Cambridge, U.S.A.: Blackwell 1991

Lightman, Bernard, ed. *Victorian Science in Context.* Chicago: University of Chicago 1997

– "'The Voices of Nature,' Popularizing Victorian Science." In Bernard Lightman, ed. *Victorian Science in Context.* Chicago: University of Chicago Press 1997

Lowenthal, David. *Possessed by the Past: The Heritage Crusade and the Spoils of History.* New York: Free Press 1996

– *The Past is a Foreign Country.* Cambridge: Cambridge University Press 1986

Lucier, Paul. "A Plea for Applied Geology." *History of Science* 36 (1999), 283–318

Mackenzie, John M., ed. *Imperialism and Popular Culture.* Manchester: Manchester University Press 1986

Macmillan, David S. *Canadian Business History: Selected Studies, 1497–1971.* Toronto: McClelland & Stewart 1972

MacRae, Marion. *Cornerstones of Order: Courthouses and Town Halls of Ontario, 1784–1914.* Toronto: Clarke, Irwin 1983

Mangan, J.A., and James Walvin, eds. *Manliness and morality: Middle-class masculinity in Britain and America 1800–1940.* New York: St. Martin's Press 1987

Marks, Lynne. "The Knights of Labor and the Salvation Army: Religion and Working Class Culture in Ontario, 1882–1890." *Labour/Le Travail,* 28 (Fall 1991), 89–127

– *Revivals and Roller Rinks: Religion, Leisure, and Identity in Late-Nineteenth-Century Small-Town Ontario.* Toronto: University of Toronto Press 1996

Marks, Shula, "History, the Nation and Empire: Sniping from the Periphery." *History Workshop Journal* 29 (Spring 1990), 111–19

Marr, William L. "The Allocation of Land to Agricultural Uses in Canada West, 1851: A View from the Individual Farm." In Donald A. Akenson, ed. *Canadian papers in rural history,* Volume X. Gananoque: Langdale Press 1996

Massey, Doreen. *Space, Place, and Gender.* Minneapolis: University of Minnesota Press 1994

– and Pat Jess, eds. *A Place in the World? Places, Culture and Globalization.* Oxford: The Open University 1995

May, Gary. *"Hard Oiler!" The Story of Early Canadians' Quest for Oil at Home and Abroad.* Toronto: Dundurn Press 1998

McBride, Robert. *Poems Satirical & Sentimental, on Many Subjects Connected with Canada, Including a Complete Exposure of our County Court and Division Court System.* London: Np. 1869

McCalla, Douglas. *Planting the Province: The Economic History of Upper Canada 1784–1870.* Toronto: University of Toronto Press 1993

McCarthy, Kathleen D. *Lady Bountiful Revisited: Women, Philanthropy and Power.* New Brunswick: Rutgers University Press 1990

McClintock, Anne. *Imperial Leather: Race, Gender and Sexuality in the Colonial Context.* New York and London: Routledge 1995

McIntosh, Robert. *Boys in the Pits: Child Labour in Coal Mines.* Montreal and Kingston: McGill-Queen's University Press 2000

McKay, Ian. *The Quest of the Folk: Antimodernism and Cultural Selection in Twentieth-Century Nova Scotia.* Montreal and Kingston: McGill-Queen's University Press 1994

McNab, David, ed. *Earth, water, air and fire: Studies in Canadian ethnohistory.* Waterloo: Wilfrid Laurier University Press 1998

Midgely, Clare, ed. *Gender and Imperialism.* Manchester and New York: Manchester University Press 1998

Miller, Orlo. *The Donnellys must die.* Toronto 1970. Originally 1962

Mills, Janice. *Crude Beginnings: A History of the Petrolia Discovery.* Petrolia: The Petrolia Discovery Foundation 2000

Morgan, Cecilia. *Public Men and Virtuous Women: The Gendered Languages of Religion and Politics in Upper Canada 1791–1850.* Toronto: University of Toronto Press 1996

Morris, R.J. *Class, sect and party: The making of the British middle class, Leeds 1820–1850.* Manchester and New York: Manchester University Press 1990

Morritt, Hope. *Rivers of Oil: The Founding of North America's Petroleum Industry.* Kingston: Quarry Press 1993

Morton, Suzanne. *At Odds: Gambling and Canadians, 1919–1964.* Toronto: University of Toronto Press 2003

Moss, Mark. *Manliness and Militarism: Educating Young Boys in Ontario for War.* Don Mills: Oxford University Press 2001

Moyles, R.G., and Doug Owram. *Imperial Dreams and Colonial Realities: British Views of Canada, 1880–1914.* Toronto: University of Toronto Press 1988

Murray, Heather. *Come, Bright Improvement! The Literary Societies of Nineteenth-Century Ontario.* Toronto: University of Toronto Press 2002

Nelles, H.V. *The Art of Nation Building: Pageantry and Spectacle at Quebec's Tercentenary.* Toronto: University of Toronto Press 1999

– *The Politics of Development: Forests, Mines & Hydro-Electric Power in Ontario, 1849–1941.* Toronto: Macmillan of Canada 1974

Newbury, Darryl Jean-Guy. "'No Atheist, Eunuch or Woman': Male Associational Culture and Working-Class Identity in Industrializing Ontario, 1840–1880." MA thesis, Queen's University, 1992

Newell, Dianne. *Technology on the Frontier: Mining in Old Ontario.* Vancouver: University of British Columbia Press 1986

O'Brien, Mike. "Manhood and the Militia Myth: Masculinity, Class and Militarism in Ontario, 1902–1914." *Labour/Le Travail* 42 (Fall 1998), 115–41

Olick, Jeffrey K., and Joyce Robbins. "Social Memory Studies: From 'Collective Memory' to the Historical Sociology of Mnemonic Practices." *Annual Review of Sociology* 24 (1998)

Ooi, Jin-Bee. *The Petroleum Resources of Indonesia.* Kuala Lumpur and New York: Oxford University Press 1982

Osborne, Brian S. "Landscapes, Memory, Monuments, and Commemoration: Putting Identity in its Place." *Canadian Ethnic Studies,* 33, 3 (2001), 39–77

Owen, Edgar Wesley. *Trek of the Oil Finders: A History of Exploration for Petroleum.* Tulsa, Oklahoma: American Association of Petroleum Geologists 1975

Palmer, Bryan D. "Historiographic Hassles: Class and Gender, Evidence and Interpretation." *Histoire sociale/Social History* 33, 65 (May 2000), 105–40

– *Working Class Experience: Rethinking the History of Canadian Labour, 1800–1991,* second edition. Toronto: McClelland & Stewart 1992

– *"A Culture in Conflict" Skilled Workers and Industrial Capitalism in Hamilton, Ontario, 1860–1914.* Montreal & Kingston: McGill-Queen's University Press 1979

Parr, Joy. "Hired Men: Ontario Agricultural Wage Labour in Historical Perspective." *Labour/Le Travail* 15 (Spring 1985), 91–103

– *The Gender of Breadwinners: Women, Men and Change in Two Industrial Towns, 1880–1950.* Toronto: University of Toronto Press 1990

Patterson, Gilbert C. *Land Settlement in Upper Canada, 1783–1840.* 16th Report of the Department of Archives, Province of Ontario, 1920.

Petrolia Canada, 1862–1908: Its Advantages as a Commercial and Residential Centre. Petrolia 1908

Phelps, Edward. "Foundations of the Canadian Oil Industry, 1850–1866." In Edith Firth, ed. *Profiles of a Province: Studies in the History of Ontario.* Toronto: Ontario Historical Society 1969

– "John Henry Fairbank of Petrolia (1831–1914): A Canadian Entrepreneur." MA thesis, University of Western Ontario, 1965

– "The Canada Oil Association – An Early Business Combination." *Western Ontario Historical Notes* 19, 2 (September 1963), 31–9

Pickering, Joseph. *Enquires of an Emigrant.* London: E. Wilson 1832

Pratt, Mary Louise. *Imperial Eyes: Travel Writing and Transculturation.* London and New York: Routledge 1992

Pratt, Wanda, and Phil Morningstar. *Early Development of Oil Technology: Oil Tools and Stories.* Oil Springs: Oil Museum of Canada 1987

Radforth, Ian. *Royal Spectacle: The 1860 Visit of the Prince of Wales to Canada and the United States.* Toronto: University of Toronto Press 2004

– *Bush Workers and Bosses: Logging in Northern Ontario 1900–1980.* Toronto: University of Toronto Press 1987

Read, Colin. "The Land Records of Old Ontario, 1791–1867." *Histoire sociale/Social History* 30, 59 (May 1997), 127–41

Reid, Fred, and David Washbrook. "Kipling, Kim and Imperialism." *History Today* 32 (August 1982), 14–20

Richards, J. Howard. "Land and Policies: Attitudes and Controls in the Alienation of Lands in Ontario During the First Century of Settlement." *Ontario History* 50, 4 (1958), 193–209

Richards, Thomas. "Archive and Utopia." *Representations* 27 (Winter 1992), 104–35

Roediger, David. *The Wages of Whiteness: Race and the Making of the American Working Class.* London: Verso 1991

Rudwick, Martin J.S. "The Emergence of a Visual Language for Geological Science 1760–1840." *History of Science* 14 (1976), 149–95

Russell, Peter A. *Attitudes to Social Structure and Mobility in Upper Canada 1815–1840: 'Here We Are Laird Ourselves'.* Lewiston, New York: E. Mellen Press 1990

Russell, Victor L., ed. *Forging a Consensus: Historical Essays on Toronto.* Toronto: University of Toronto Press 1984

Ryan, Mary P. *Cradle of the Middle Class: The Family in Oneida County, New York, 1790–1865.* New York: Cambridge University Press 1981

– *Women in Public: Between Banners and Ballots, 1825–1880.* Baltimore and London: Johns Hopkins University Press 1990

Saddlemeyer, Ann. ed. *Early Theatre in Ontario 1800–1914.* Toronto: University of Toronto Press 1990

Said, Edward W. *Orientalism.* New York: Vintage Books 1979

Samson, Daniel Joseph. "Industry and Improvement: State and Class Formations in Nova Scotia's Coal-Mining Countryside, 1790–1864." PhD dissertation, Queen's University 1997

Samuel, Raphael. *Theatres of Memory, Vol. 1: Past and Present in Contemporary Culture.* London and New York: Verso 1994. Reprinted 1996

Scott, Joan Wallach, ed. *Feminism and History*. Oxford and New York: Oxford University Press 1996

Scranton, Philip, ed. *Beauty and Business: Commerce, Gender, and Culture in Modern America*. New York: Routledge 2001

Secord, Anne. "Corresponding interests: artisans and gentlemen in nineteenth-century natural history." *British Journal of the History of Science* 27 (1994), 283–408

– "Science in the Pub: Artisan Botanists in Early Nineteenth-Century Lancashire." *History of Science* 32 (1994), 269–315

Secord, James A. *Controversy in Victorian Geology: The Cambrian-Silurian Dispute*. Princeton, N.J.: Princeton University Press 1986

– "The Geological Survey of Great Britain as a Research School, 1839–1855." *History of Science* 24 (1986), 223–75

– *Victorian Sensation: The Extraordinary Publication, Reception, and Secret Authorship of Vestiges of the Natural History of Creation*. Chicago and London: University of Chicago Press 2000

Shannon, William. "Brokers, Land Bankers, and 'Birds of Evil Omen': The Effect of Land Policies on Settlement in Upper Canada Collingwood Township 1834–1860." MA thesis, University of Ottawa, 1989

Shaw, Gareth. "Nineteenth Century Directories as Sources in Canadian Social History." *Archivaria* 14 (Summer 1982), 107–21

Shields, Rob. *Places on the Margin: Alternative geographies of modernity*. London and New York: Routledge Chapman Hall 1991

Simcoe, Elizabeth. *The Diary of Mrs John Graves Simcoe, Wife of the First Lieutenant Governor of the Province of British Columbia, 1792–6*. Toronto: W. Biggs 1911

Smith, Charles H. "William Logan's 1850 History of the Geological Survey of Canada." *Geoscience Canada* 26, 3 (September 1999): 11–20

Soja, Edward W. *Postmodern Geographies: The Reassertion of Space in Critical Social Theory*. London: Verso 1989

Spurr, David. *The Rhetoric of Empire: Colonial Discourse in Journalism, Travel Writing, and Imperial Administration*. Durham: Duke University Press 1993

Stafford, Robert A. *Scientist of empire: Sir Roderick Murchison, scientific exploration and Victorian imperialism*. Cambridge: Cambridge University Press 1989

Stasiulis, Daiva, and Uira Yuval-Davis. *Unsettling Settler Societies: Articulations of Gender, Race, Ethnicity and Class*. London: Sage 1995

Stoler, Ann Laura. "Carnal Knowledge and Imperial Power: Gender, Race, and Morality in Colonial Asia." In Joan Wallach Scott, ed. *Feminism and History*. Oxford and New York: Oxford University Press 1996

– "Rethinking Colonial Categories: European Communities and the Boundaries of Rule," *Comparative Studies in Society and History*, 13 (1989), 134–61

– *Capitalism and Confrontation in Sumatra's Plantation Belt, 1870–1979*. New Haven: Yale University Press 1985

Stowe, William W. *Going Abroad: European Travel in Nineteenth-Century American Culture*. Princeton: Princeton University Press 1994

Surtees, Robert J. "Indian Land Cessions in Upper Canada, 1814–1830." In Ian A.L. Getty and Antoine S. Lussier, eds. *As Long as the Sun Shines and the Water Flows*. Vancouver: University Of British Columbia Press 1983

Sutherland, David. "Voluntary Societies and the Process of Middle-class Formation in Early-Victorian Halifax, Nova Scotia." *Journal of the Canadian Historical Association* (1994), 237–63

Tabili, Laura. *"We Ask for British Justice": Workers and Racial Difference in Late Imperial Britain*. Ithaca and London: Cornell University Press 1994

Tallentire, Jenéa. "Strategies of Memory: History, Social Memory, and the Community." *Histoire sociale/Social History* 34, 67 (May 2001), 197–212

Teich, Mikuláš, and Robert Young, eds. *Changing Perspectives in the History of Science*. Dordrecht and Boston: D. Reidel 1973

Telford, Rhonda. "'Under the Earth': The Expropriation and Attempted Sale of the Oil and Gas Rights of the Walpole Island First Nation During World War I." In David McNab, ed. *Earth, water, air and fire: Studies in Canadian ethnohistory*. Waterloo: Wilfrid Laurier University Press 1998

Thelen, David. "Memory and American History." *The Journal of American History* 75, 4 (March 1989), 1122–3

Thrift, Nigel, and Peter Williams, eds. *Class and space: The making of urban society*. London and New York: Routledge & Kegan Paul 1987

Torrens, H.S. "William Edmond Logan's Geological Apprenticeship in Britain 1831–1842." *Geoscience Canada* 26, 3 (September 1999), 97–110

Traill, Catharine Parr. *The Canadian Settler's Guide*. Toronto 1969. Originally 1855

Urry, John. *The Tourist Gaze: Leisure and Travel in Contemporary Societies*. London: Sage Publications 1990

Vernon, Foster. "The Development of Adult Education in Ontario 1790–1900." PhD thesis, University of Toronto, 1969

Voisey, Paul. *Vulcan: The Making of a Prairie Community*. Toronto: University of Toronto Press 1988

Walden, Keith. *Becoming Modern in Toronto: the Industrial Exhibition and the Shaping of a Late Victorian Culture*. Toronto: University of Toronto Press 1997

Wallace, Mike. *Mickey Mouse History and Other Essays on American Memory*. Philadelphia: Temple University Press 1996

Walsh, John C. "Property, Surveyors, and State Formation." Unpublished paper presented to the Annual Meeting of the Canadian Historical Association, Toronto 2002

Walsh, John C., and Steven High. "Rethinking the Concept of Community." *Histoire sociale/Social History* 32, 64 (November 1999), 255–73

Walsh, Kevin. *The Representation of the Past: Museums and heritage in the post-modern world.* London and New York: Routledge 1992

Wamsley, Kevin B. "The Public Importance of Men and the Importance of Public Men: Sport and Masculinities in Nineteenth-Century Canada." In Philip White and Kevin Young, eds. *Sport and Gender in Canada.* Oxford: Oxford University Press 1999

Whipp, Charles, and Edward Phelps. *Petrolia 1866–1966.* Petrolia 1966

White, Henry. *Geology, Oil Fields and Minerals of Canada West: How and Where to Find Them.* Toronto: W.C. Chewett 1865

White, Philip, and Kevin Young, eds. *Sport and Gender in Canada.* Oxford: Oxford University Press 1999

Widdis, Randy William. "A Perspective on Land Tenure in Upper Canada: A Study of Elizabethtown Township, 1790–1840." MA thesis, McMaster University, 1977

– "Motivation and Scale: A Method of Identifying Land Speculators in Upper Canada." *Canadian Geographer* 23, 4 (1979), 338–51

Williams, Raymond. *The Country and the City.* New York: Oxford University Press 1973

Winchell, Alexander. *Sketches of Creation. A Popular View of Some of the Grand Conclusions of the Sciences in Reference to the History of Matter and of Life.* New York: Harper 1873

Wise, S.F. "Sport and Class Values in Old Ontario and Quebec." In W.H. Heick and Roger Graham, eds. *His Own Man: Essays in Honour of Arthur Reginald Marsden Lower.* Montreal: McGill-Queen's University Press 1974

Zaslow, Morris. *Reading the Rocks: The Story of the Geological Survey of Canada.* Ottawa: Macmillan 1975

Zeller, Suzanne. *Inventing Canada: Early Victorian Science and the Idea of a Transcontinental Nation.* Toronto: University of Toronto Press 1987

Index